RISK

Frontispiece *"Newton" by William Blake (source:* Tate Gallery, London/Bridgeman Art Library, London).

RISK

John Adams
University College London

London and New York

First published in 1995 by UCL Press
Third impression 1996
Fouth impression 1998
Fifth impression 2000

First published 2001 by Routledge
11 New Fetter Lane
London EC4P 4EE

Routledge is an imprint of the Taylor & Francis Group

British Library Cataloguing in Publication Data
A catalogue record for this book is available from the British Library.

Library of Congress Cataloguing in Publication Data
Adams, John, 1938–
    Risk : the policy implications of risk compensation and plural
rationalities / John Adams.
    p. cm.
    Includes bibliographical references and index.
    ISBN 1–85728–067–9. — ISBN 1–85728–068–7 (pbk.)
    1. Risk—Sociological aspects. 2. Risk management—Social
aspects. I. Title.
HM256.A33 1995
302´.12—dc20   95–88
                                                CIP

ISBNs: 1–85728–067–9 HB
         1–85728–068–7 PB

Typeset in Zapf Elliptical.
Printed and bound by
Biddles Ltd, Guildford and King's Lynn, England

CONTENTS

v

PREFACE

This book began as a collaborative venture with Michael Thompson. For over 15 years my research into risk, mainly on the road, was focused on the theory of "risk compensation". This theory accords primacy in the explanation of accidents to the human propensity to take risks. The theory postulates that we all come equipped with "risk thermostats" and suggests that safety interventions that do not affect the setting of the thermostat are likely to be frustrated by behavioural responses that reassert the level of risk with which people were originally content. My research had noted that there were large variations in the settings of individual thermostats, but had little to say about why this should be so.

About ten years ago I read Michael's article "Aesthetics of risk" (Thompson 1980), and about five years later met the man himself. His research into risk over the past 20 years has been central to the development of a perspective that has come to be known as "cultural theory" (Thompson et al. 1990). Risk, according to this perspective, is *culturally constructed*; where scientific fact falls short of certainty we are guided by assumption, inference and belief. In such circumstances the deterministic rationality of classical physics is replaced by a set of conditional, probabilistic rationalities. Risk throws up questions to which there can be no verifiable single right answers derivable by means of a unique rationality. Cultural theory *illuminates a world of plural rationalities*; it discerns order and pattern in risk-taking behaviour, and the beliefs that underpin it. Wherever debates about risk are prolonged and unresolved – as, for example, that between environmentalists and the nuclear industry – cultural theory seeks an explanation not in further scientific analysis but in the differences in premises from which the participants are arguing. Michael thought that risk compensation was obvious common sense, and I thought that cultural theory would cast helpful light on how the thermostat was set.

This book grew out of a joint research project called "Risk and rationality" that we undertook for the Economic and Social Research Council. It draws upon much of our earlier work, and makes connections that had earlier eluded us. When we first discussed the idea of a book with Roger Jones of UCL Press we rashly promised to produce "the complete theory of risk".

Trying has been an educational experience, but the complete theory of risk now seems to me as likely as the complete theory of happiness.

The writing did not go as planned – but then this is a book about risk. Michael, who is self-employed, was distracted by consultancy offers from all around the world that he could not refuse. I stayed at home and got on with the writing, making use of Michael, when I could catch him, as a consultant. Inevitably the book does not have the balance originally intended between his perspective and mine. In Chapter 11 I refer to the "tension" between cultural theory and risk compensation. This refers to my unresolved difficulty in reconciling cultural theory with the *reflexivity* of risk. The world and our perceptions of it are *constantly* being transformed by our effect on the world, and its effect on us. My perceptions of risk have been altered by the process of writing this book. I now see the stereotypes of cultural theory – egalitarians, individualists, hierarchists, fatalists and hermits – everywhere I look. But which am I?

I think I can see elements of all these stereotypes in my own make up. Am I more sophisticated and complex than all the other risk-takers in the world? I doubt it. In applying these stereotypes to others I am reducing their complex uniqueness to something that I can (mis)understand. In its raw state the reflexive fluidity of the world overwhelms our limited powers of comprehension. We resort to simplification and abstraction in an attempt to cope. Cultural theory postulates a high degree of pattern and consistency in the midst of all the reflexive fluidity. The insistence in *cultural theory* (Thompson 1990) on the *impossibility* of more than five "viable ways of life" I find unproven and unprovable, but I still find the theory useful. For me, limiting the number of risk-taking types to five is defensible, not just by theoretical speculation, but by virtue of five being a small and comprehensible number; theories of behaviour, to be useful and widely communicable, must be simple. Risk compensation and cultural theory provide a life-raft that saves one from drowning in the sea of reflexive relativism; they are two sets of simplifying assumptions deployed in this book in an attempt to make sense of behaviour in the face of uncertainty. They are not the complete theory of risk.

In "test marketing" draft chapters of the book on a variety of people with an interest in risk, it became apparent that many from the scientific and managerial side of the subject are unaware of the anthropological literature on risk, and its roots in the work of Weber, Durkheim, Marx, Malinowksi, Parsons and other old masters of sociology and anthropology; they have reacted with scepticism and impatience to the theorizing of Douglas, Wildavsky, Thompson and other, more recent, workers in this tradition. On the other hand, some in this tradition have complained that my treatment of cultural theory is "superficial and derivative" – to quote from the comments of one referee on a part of Chapter 3 which was submitted to an academic journal as an article. The literature on risk, measured by pages published, is overwhelmingly dominated by the scientific/managerial perspective. In trying to make cultural theory accessible to the scientist–managers, I have stripped it of most of its historical baggage, and many of its claims to "scientific" authority. I have retained what I consider to be its easily communicated essence; I have treated it as a set of abstractions that help to make sense of many interminable debates about risk. I have no illusions that my efforts to bridge the divide between the "hard" and "soft" approaches to risk will satisfy everyone – indeed cultural theory warns that everyone will never agree about risk. But attempting the impossible has been fun.

John Adams
LONDON

ACKNOWLEDGEMENTS

By far the largest debt incurred in writing this book is owed to Michael Thompson. His wide knowledge of the anthropological literature, his shrewd insights, his gift for seeing a problem from a new angle, his patience when he was having difficulty getting information through my cultural filter (see Ch. 3 on the subject of cultural filters), and above all his ability to disagree agreeably, have made the writing an enormously educational, stimulating and enjoyable experience. I look forward to arguing with him for years to come.

The earliest, indeed formative, influence on my thoughts about risk was Gerald Wilde, who coined the term "risk compensation". I thank him for his hospitality and many entertaining tutorials on the subject over the years. As I observe in Chapter 2, *everyone* is a risk expert. This has made the job of consulting the experts quantitatively daunting. Argument, I believe, is the most educational form of discourse, and this book is the result of years of arguing with just about anyone who would tolerate my banging on about the subject; for the risk researcher, life is one never-ending field trip. This makes the task of acknowledging all my debts quite impossible – I did not always make notes at the time – but toward the end of the process, the participants in our ESRC-sponsored workshop on risk and rationality helped me to get my thoughts into focus: David Ball, David Collingridge, Karl Dake, Mary Douglas, Maurice Frankel, Gunnar Grendstadt, Joanne Linnerooth-Bayer, Mark MacCarthy, Gustav Östberg, Alex Trisoglio, Brian Wynne. In addition Bob Davis, John Whitelegg, Mayer Hillman, Stephen Plowden, Robin Grove White, Edmund Hambly, Jacquie Burgess, Carolyn Harrison, and the late Aaron Wildavsky, have all been helpful. I doubt that any of these people would agree with all of this book but, whether they like it or not, they have all had an influence on it.

Louise Dyett, Tim Aspden and Guy Baker in the UCL Department of Geography drawing office have played a vital rôle in producing, and helping to design, the illustrations.

Anna Whitworth's constructive criticism of preliminary drafts has been much appreciated. I am also grateful for her sharp editorial eye which has prevented the publication of many spelling mistakes and lapses in political correctness.

Chapter 1

RISK:
AN INTRODUCTION

One of the pleasures of writing a book about risk – as distinct from one about an esoteric subject such as brain surgery or nuclear physics – is that one has a conversation starter for all occasions. *Everyone* is a true risk "expert" in the original sense of the word; we have all been trained by practice and experience in the management of risk. *Everyone* has a valid contribution to make to a discussion of the subject.

The development of our expertise in coping with uncertainty begins in infancy. The trial and error processes by which we first learn to crawl, and then walk and talk, involve decision-making in the face of uncertainty. In our development to maturity we progressively refine our risk-taking skills; we learn how to handle sharp things and hot things, how to ride a bicycle and cross the street, how to communicate our needs and wants, how to read the moods of others, how to stay out of trouble. How to stay out of trouble? This is one skill we never master completely. It appears to be a skill that we do not want to master *completely*.

The behaviour of young children, driven by curiosity and a need for excitement, yet curbed by their sense of danger, suggests that these junior risk experts are performing a balancing act. In some cases it is a physical balancing act; learning to walk or ride a bicycle cannot be done without accident. In mastering such skills they are not seeking a zero-risk life; they are balancing the expected rewards of their actions against the perceived costs of failure. The apprehension, determination and intense concentration that can be observed in the face of a toddler learning to toddle, the wails of frustration or pain if it goes wrong, and the beaming delight when it succeeds – are all evidence that one is in the presence of a serious risk-management exercise.

Most decisions about risks involving infants and young children are taken by adults. Between infancy and adulthood there is a progressive handing over of responsibility. Adults are considered *responsible* for their actions, but they are not always considered trustworthy or sufficiently well informed. A third tier of responsibility for the management of risk consists of various *authorities* whose rôle with respect to adults is similar to that of adults with

1

respect to children. The authorities are expected to be possessed of superior wisdom about the nature of risks and how to manage them.

The news media are routinely full of stories in which judgement is passed on how well or badly this expectation is met. Consider an ordinary news day chosen at random – 28 January 1994, the day this sentence was written. A perusal of that day's papers[1] reveals that the business sections and the sports pages contain virtually no stories that are not about the management of risk. They are all about winning and losing, and winners and losers. The heroes are people who struggled against the odds and won. Prudence and caution, except for the occasional bit of investment advice for old age pensioners, are mocked as boring. The arts pages were full of risk stories within risk stories. A novel, to win critical acclaim, must be *novel*; cliché and plagiarism are unpardonable sins. Mere technical competence is not enough; suspense and tension must be deployed to catch and hold the attention of the reader. Risk is embodied in great works of art; and, to capture the interest of the arts pages, risks must be taken by their creators. They are interesting only if they are attempting something difficult. Great art risks failure. But to be boring, predictable and safe is to guarantee failure.

What of the features pages? The motoring sections of most of the papers were dominated as usual by articles focused on the performance of cars – although the main feature in one was devoted to question of whether or not airbags caused injuries, and another paper ran a small story about a new car seat for children, with the claim that it "reduced the risk by 90%". The lifestyle section of another ran a double-page spread on high-performance motorcycles under the headline "Born to be wild".

The health pages were of course entirely devoted to risk stories: a new chickenpox vaccine whose effectiveness remains to be proven; a series of mistakes in cervical cancer screening that "put patients' lives at risk"; the risk of blood transfusions transmitting hepatitis-B; a vasectomy that did not work; concern that epidural anaesthetics administered during childbirth might harm the babies; the fear that bovine spongiform encephalopathy might have spread to humans in the form of Creutzfeld-Jakob disease; doubts about the efficacy of drugs prescribed to control high blood pressure; doubts about the accuracy of the diagnosis of high blood pressure, and claims that it is increased by the act of measuring it; claims that "the Government's present [health] screening programme cannot be justified by the results"; a lottery held to choose who would be given a scarce new and unproven drug for treating multiple sclerosis; and a member of parliament who died while waiting for a heart transplant, with credit for the shortage of donors being given to the seat belt law. Even the gardening pages were dominated by problems of decision-making in the face of uncertainty: combinations of soil,

---

1. *The Times*, the *Guardian*, the *Sun*, the *Daily Express*, the *Daily Mail*, and the London *Evening Standard*.

climate, aspect, fungicides and insecticides *might* be propitious for this plant and not for that.

The news pages were overwhelmingly devoted to risk. Risk it would appear is a defining characteristic of "news". On 28 January 1994 an aid worker had been killed in Bosnia; the US President's wife, Hilary Clinton, visited the aftermath of the Los Angeles earthquake, most of whose victims were reported to be uninsured; an Englishman staked his life savings of £150,000 on one spin of the roulette wheel in Las Vegas, and won; the death of a budgerigar was blamed on passive smoking, and a woman was turned down as a prospective adoptive parent because she smoked; the roof of a supermarket in Nice collapsed killing three people (56 column-inches), and a fire in a mine in India killed 55 people (nine column-inches); Prince Charles was fired at in Australia by a man with a starting pistol, and Princess Diana's lack of security was highly publicized, and lamented; further restraints were threatened on cigarette advertising; death threats were made by Moslem fundamentalists to a couturier and a fashion model following publicity about a ball gown embroidered with a passage from the Koran; the Government launched its "green" plan, and environmentalists complained about its in-adequacy. A few more headlines: "Rogue train ignored signals", "Russia's high-risk roulette", "Mountaineer cleared of blame for woman's death fall", "£440,000 losers in a game of Russian roulette (the costs of a lost libel action)", "Libel law proves costly lottery", "Fall in family fortunes", "The cat with 11 lives", "Gales strand trains and cause road havoc", "Fire-bombs in Oxford St raise fear of fresh IRA campaign", "Israelis have 200 N-bombs" and "Diet-conscious add years to life expectancy".

Television news and documentary programmes on the same day provided a further generous helping of things to worry about, and films added fictional accounts of neurosis, angst, murder and mayhem. Daily we are confronted with a fresh deluge of evidence that in this world nothing can be said to be certain, except death – stories of large-scale tax evasion having removed taxes from the short list of certainties. How do we cope?

Grown-up risk-taking, like that of children, is a balancing act. Whether it be the driver at the wheel of a car negotiating a bend in an icy road, or a shopper trying to decide whether to buy butter or the low-fat spread, or a doctor trying to decide whether to prescribe a medicine with unpleasant side-effects, or a property speculator contemplating a sale or a purchase, or a general committing his troops to battle, or a President committing his country to curbing the emission of carbon dioxide, the decisions that are made in the face of uncertainty involve weighing the potential rewards of an act against its potential adverse consequences.

Every day around the world, billions of such decisions get made. The consequences in most cases appear to be highly localized, but perhaps they are not. Chaos theorists have introduced us to a new form of insect life called the Beijing butterfly – which flaps its wings in Beijing and sets in motion a train of events that culminates two weeks later in a hurricane in New York. Extreme sensitivity to subtle differences in initial conditions, the chaos theorists tell us, makes the behaviour of complex natural systems inherently unpredictable. Prediction becomes even more difficult when *people* are introduced to such systems – because people respond to predictions, thereby altering the predicted outcome. Rarely are risk decisions made with information that can be reduced to quantifiable probabilities; yet decisions, somehow, get made.

The universality of expertise in risk management is a problem for those who aspire to recognition as risk *EXPERTS*. The certified experts – those who write books, learned articles and official reports on risk – have an abstracted expertise that is sometimes useful, but is more often misleading. They can demonstrate that the general public's ability to estimate mortality rates for different causes of death is often very wide of the mark (Fischhoff et al. 1981); they can demonstrate, in the words of the Royal Society quoted in Chapter 2, that there is a "gap between what is scientific and capable of being measured, and the way in which public opinion gauges risks and makes decisions". They can demonstrate that ordinary people in managing the risks in their lives, rarely resort to precise quantification. But what do their scientific measurements signify? Very little, this book suggests.

Risk management is big business; the *formal sector* of the authorities – the realm of the expert – involves government, commerce, and industry; it employs actuaries, ambulance drivers, toxicologists, engineers, policemen, mathematicians, statisticians, economists, chaos theorists, computer programmers and driving instructors – to name but a few. The work of this sector is highly visible. It holds inquests and commissions research. It passes laws and formulates regulations. It runs safety training programmes and posts warning signs. It puts up fences and locks gates. It employs inspectors and enforcers – many in uniform. Its objective is to *reduce* risk.

But there is also the *informal sector* consisting of children and grown-up children, and it is much bigger business; it consists of billions of freelance risk managers – ordinary common-or-garden experts – each with his or her own personal agenda. They go about the business of life – eating, drinking, loving, hating, walking, driving, saving, investing, working, socializing – striving for health, wealth and happiness in a world they know to be uncertain. The objective of these risk managers is to *balance* risks and rewards.

The formal and informal sectors co-exist uncomfortably. For the freelance risk managers, the activities of the formal sector form a part of the context within which they take their decisions. Sometimes the efforts of the formal sector are appreciated: when, for example, it assumes responsibility for the safety of the water you drink. Sometimes its efforts are thought to be inadequate: when it fails to slow down the traffic on your busy street. Sometimes its efforts are resented: when it sets speed limits too low, or its safety regulations interfere with activities you consider safe enough. But in all of these cases, behaviour in the informal sector is modified by the activities of the formal sector. You do not boil your water if they have made it safe. You take more care crossing the road that their negligence makes dangerous. You watch out for the police or safety inspectors whose silly rules you are breaking.

The formal sector responds to the activities of freelance risk-managers in various ways. Often it is patronizing. Road engineers with their accident statistics frequently dismiss condescendingly the fears of people living alongside busy roads with good accident records, heedless of the likelihood that the good accident records reflect the careful behaviour of people who believe their roads to be dangerous. Those who live alongside such roads, and know their dangers, are more likely than the engineer, beguiled by his statistics, to cross them safely. Sometimes the formal sector's response is abusive: the people who flout their rules are stupid, irresponsible or childish. But most commonly the formal sector is mystified and frustrated. How, they wonder – despite all their road improvements, vehicle safety regulations, speed limits, alcohol limits, warning notices, inspection procedures and fail-safe devices – do so many people still manage to have accidents?

A significant part of the explanation appears to lie in the formal sector's division of labour. Risk-management at an individual level involves no division of labour; the balancing calculations that precede a risky act are all done in the head of the individual. But when institutions assume responsibility for risk management, it becomes difficult to identify where the balancing act is done. Consider road safety. One can list institutions concerned with maximizing the rewards of risk taking: the car industry, the oil industry, the road builders, that part of the Department of Transport which sees its function as aiding and abetting the process that generates increasing traffic, the Treasury and Department of Trade and Industry who point to this increase as evidence of growing prosperity. One can list other institutions concerned with minimizing the accident costs of road traffic: the police, the casualty services, PACTS (the Parliamentary Advisory Committee on Traffic Safety), ROSPA (the Royal Society for the Prevention of Accidents), Friends of the Earth and Greenpeace, who are concerned about the global threats of traffic pollution as well as the danger to cyclists and pedestrians, and that part of the Department of Transport responsible for road safety. But where, and how, is the balancing act done? How do institutional risk-managers manage individual risk-managers? And how do individual risk-managers react to attempts to manage them? And can we all do it better?

The search for answers begins, in Chapter 2, with a look at the prevailing orthodoxy, as exemplified by the Royal Society's approach to risk management.

Chapter 2

RISK AND
THE ROYAL SOCIETY

In 1983 Britain's Royal Society published a report called *Risk assessment*. Its tone, in keeping with the Royal Society's standing as the country's pre-eminent scientific institution, was authoritative, confident and purposeful. The report drew upon and exemplified the prevailing international ortho-doxy on the subject of risk, and became a major work of reference. In 1992 the Society returned to the subject with a new report entitled *Risk: analysis, perception and management*. Although it was published by the Royal Soci-ety, the Society was sufficiently embarrassed by its contents to insist in the preface that it was "not a report of the Society", that "the views expressed are those of the authors alone" and that it was merely "a contribution to the ongoing debate". By 1992 the Royal Society was no longer capable of taking a collective view about risk; it had become a mere forum for debate about the subject. What happened? What is this "ongoing debate", and how did it derail their inquiries into the subject?

For their 1992 report the Society invited a group of social scientists – psychologists, anthropologists, sociologists, economists and geographers – to participate in their study. The social scientists, with the exception of the economists, could not agree with the physical scientists of the Royal Society. The disagreement that is found between the covers of the 1992 report can be found wherever there are disputes about safety and danger. It is a dis-agreement about the nature and meaning of "risk". The resolution of this disagreement will have profound implications for the control and distribu-tion of risk in all our lives.

"Actual risk": what is it?

The 1983 report distinguished between *objective risk* – the sort of thing "the experts" know about – and *perceived risk* – the lay person's often very different anticipation of future events. Not surprisingly, given the report's provenance, it approached its subject *scientifically*. This is how it defined the subject of its study in 1983:

> The Study Group views "risk" as the probability that a particular adverse event occurs during a stated period of time, or results from a particular challenge. As a probability in the sense of statistical theory, risk obeys all the formal laws of combining probabilities.

The Study Group also defined *detriment* as:

> a numerical measure of the expected harm or loss associated with an adverse event . . . it is generally the integrated product of risk and harm and is often expressed in terms such as costs in £s, loss in expected years of life or loss of productivity, and is needed for numerical exercises such as cost–benefit analysis or risk–benefit analysis.

The Royal Society's definition of "detriment", as a compound measure combining the probability and magnitude of an adverse effect, is the definition of "risk" most commonly encountered in the risk and safety literature (see for example Lowrance 1980). It is also the definition of common parlance; people do talk of the "risk" (probability) of some particular event being high or low, but in considering two possible events with equal probabilities – say a fatal car crash and a bent bumper – the former would invariably be described as the greater risk. But, definitional quibbles aside, the Royal Society and most other contributors to the risk literature are agreed about the objective nature of the thing they are studying. There is also general agreement that progress lies in doing more of what physical scientists are good at: refining their methods of measurement and collecting more data on both the probabilities of adverse events and their magnitudes. One of the main conclusions of the 1983 report was that there was a need for "better estimates of actual risk based on direct observation of what happens in society" (p. 18).

Across the Atlantic in 1983 the American scientific establishment was also taking an interest in risk and coming to similar conclusions. The National Research Council, which is the principal operating agency for the National Academy of Sciences and the National Academy of Engineering, published a report entitled *Risk assessment in the Federal Government: managing the process*. Like their counterparts in the Royal Society, they stressed the importance of the distinction between the "*scientific basis*" and the "*policy basis*" of decisions about risk. Their report repeatedly stressed the importance of maintaining "a clear conceptual distinction between assessment of risks and the consideration of risk management alternatives." They warned that "even the *perception* that risk management considerations are influencing the conduct of risk assessment in an important way will cause the assessment and regulatory decisions based on them to lack credibility".

In the study leading up to its 1992 report, the Royal Society set out maintaining the distinction between objective and perceived risk. The Study Group's terms of reference invited it to:

> consider and help to bridge the gap between what is stated to be scientific and capable of being measured, and the way in which public opinion gauges risks and makes decisions.

It failed. The gap remains unbridged. The introduction of the 1992 report repeats the 1983 report's definitions of risk and detriment, and the first four chapters of its 1992 publication still cling to the distinction between objective and perceived risk. They are illustrated by the usual tables of objective risk – the risk of dying from being a miner, or a smoker, or not wearing a seat belt, and so on. They contain many qualifications about the accuracy of many risk estimates, and admonitions against using them mechanistically, but these warnings are presented as exhortations to try harder to obtain accurate quantified estimates of objective risk. Chapter 4 concludes that "if risk assessment is to be more than an academic exercise, it must provide quantitative information that aids decisions . . .".

However, by Chapter 5 the distinction between objective risk and perceived risk, fundamental to the approach of the Royal Society's 1983 report and the first four chapters of its 1992 report, is flatly contradicted:

> the view that a separation can be maintained between "objective" risk and "subjective" or perceived risk has come under increasing attack, to the extent that it is no longer a mainstream position.

A contention of chapters 5 and 6 of the 1993 report, that the physical scientists found variously maddening or frustrating, is that risk is culturally constructed. According to this perspective, both the adverse nature of particular events and their probability are inherently subjective. Slipping and falling on the ice, for example, is a game for young children, but a potentially fatal accident for an old person. And the probability of such an event is influenced both by a person's perception of the probability, and by whether they see it as fun or dangerous. For example, because old people see the risk of slipping on an icy road to be high, they take avoiding action, thereby reducing the probability. Young people slipping and sliding on the ice, and old people striving to avoid doing the same, belong to separate and distinct *cultures*. They *construct* reality out of their experience of it. They see the world differently and behave differently; they tend to associate with kindred spirits, who reinforce their distinctive perspectives on reality in general and risk in particular.

Before exploring (in Ch. 3) the variety of ways in which risk is constructed, I turn first to the way in which this variety frustrates those who seek to subject risk to the measuring instruments of objective science.

Can risk be measured?

Lord Kelvin once said "Anything that exists, exists in some quantity and can therefore be measured" (quoted in Beer 1967).

Physical scientists tend to be suspicious of phenomena whose existence cannot be verified by objective replicable measurement, and Kelvin's dictum epitomizes the stance of those who address themselves to the subject of risk. They might be called "objectivists", or perhaps "Kelvinists" in keeping with the theological character of their position – the dictum that underpins their objective science is itself incapable of objective proof.

Chapter 5 of the 1992 Royal Society report overstates the current strength of the opposition to the Kelvinists. The view that there is a distinction to be made between, *real, actual, objective, measurable risk that obeys the formal laws of statistical theory* and *subjective risk inaccurately perceived by non-experts* is still the mainstream position in most of the research and literature on safety and risk management. Certainly the view that there is no such thing as "objective risk" and that risk is "culturally constructed" is one that some members of the Royal Society appear to find incomprehensible, and others, more robustly, dismiss as relativistic, airy-fairy nonsense. Much can depend on whether or not they are right.

Britain's Department of Transport belongs to the Kelvinist camp. It measures the safety or danger of a road by its casualty record – the consequences of real accidents. It draws a clear line between actual danger and perceived danger. The Department is prepared to spend money only to relieve actual danger. If a road does not have a fatality rate significantly above "normal" (about 1.2 per 100 million vehicle kilometres), it is not eligible for funds for measures to reduce the danger.

Sir Patrick Brown (1991), Permanent Secretary of Britain's Department of Transport, has announced that "funds for traffic calming will be judged on casualty savings, not environmental improvements or anxiety relief". All up and down the country there are people living alongside roads that they perceive to be dangerous, but which have good accident records. They are told in effect that if you don't have blood on the road to prove it, your road is officially, objectively, safe, and your anxiety is subjective and emotional.

In the road safety literature, and the safety literature generally, it is still the mainstream position that casualty statistics provide the only reliable measure of the success or failure of safety schemes. The foremost academic journal devoted to safety issues is *Accident analysis and prevention*; the metric of success and failure is embedded in its title. Safe roads, or airlines, or factories, or hospitals, or schools, or playgrounds, are those with few accidents. The objective of accident analysis is accident prevention.

Why was the Royal Society studying risk in the first place? In 1983 they put it this way:

Governments are now seen to have a plain duty to apply themselves explicitly to making the environment safe, to remove *all* risk or as much of it as is reasonably possible.

They were seeking to offer advice about how risk might be eliminated, or reduced, or at least better managed. By 1992 the objective of managing risk had been included in the title of their report *Risk: analysis, perception and management*. This, one might think, is a worthwhile and uncontentious objective; it is one shared with hundreds of journals and campaigning organizations concerned with safety all around the world. The "plain duty" to reduce accidents permeates the study of risk. If risk exists, according to the Kelvinists, it exists as a probability that can be measured – by accident statistics. But can it?

The area of risk-taking that generates the greatest volume of accident statistics is danger on the road. It is a category of risk that can be clearly distinguished from other areas of risk-taking activity. Although major problems are encountered in defining the categories of injury severity, the *fatality* statistics in most highly motorized countries are probably accurate within a few percentage points, and the circumstances of each fatal accident are recorded systematically and in considerable detail. Furthermore, the numbers of fatal accidents are large enough to permit statistical analysis of intervention effects. The British Medical Association (1980) has observed that

deaths and injuries on the road are one of the few subjects where preventive medicine can be based on reliable statistics on the effects of intervention.

But controversy still surrounds the interpretation of these statistics. Consider this view of the change over time of safety on the roads.

I can remember very clearly the journeys I made to and from school because they were so tremendously exciting . . . The excitement centred around my new tricycle. I rode to school on it every day with my eldest sister riding on hers. No grown-ups came with us . . . All this, you must realize, was in the good old days when the sight of a motor car on the street was an event, and it was quite safe for tiny children to go tricycling and whooping their way to school in the centre of the highway. (Roald Dahl, in *Boy*, recalling his childhood in Llandaff, Glamorgan in 1922.)

The young Roald Dahl was doing something that was "tremendously exciting" and yet "quite safe". Was he taking a risk?

Figure 2.1 shows that between 1922 and 1986, while the motor vehicle population of England and Wales increased 25 fold, the number of children under the age of 15 killed in road traffic accidents fell from 736 per annum to 358. Allowing for changes in population, the road accident death rate for children is now about half its level of 70 years ago. The child road death rate per motor vehicle has fallen by about 98 per cent.

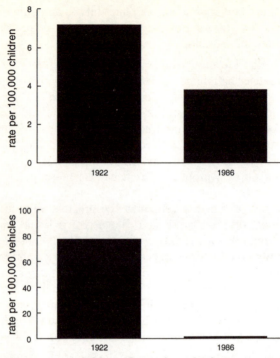

Figure 2.1 Child road accident fatalities (*source:* Hillman et al. 1990).

On the basis of these statistics, the conventional wisdom would conclude that Roald Dahl's *subjective* belief that the roads used to be "quite safe" was simply wrong; *objectively* – statistically – they have become very much safer. Certainly, this is the way politicians and safety officials routinely interpret the evidence. A British Government road safety report (Department of Transport 1990) says, for example:

> Over the last quarter of a century, Britain's roads have become much safer. Road accidents have fallen by almost 20% since the mid-1960s; the number of deaths is down by one third. At the same time, traffic has more than doubled.

The orthodox school of risk assessment treats accident statistics as objective measures of risk. The most commonly used scale is the number of "events" (accidents or injuries) per 100,000 persons per unit of time (see, for example, Urquhart & Heilmann 1984 and Hambly 1994). These measures are interpreted by *experts* as objective indices of risk, and sometimes compared with the subjective judgements of lay people, usually with the aim of demonstrating the hopeless inaccuracy of the latter. A British Medical Association study (1987), for example, reports enormous disparities between lay estimates of deaths and actual deaths attributed to a variety of causes. Highway engineers swap anecdotes about roads that have good accident records, but which nevertheless provoke complaints from local residents about their danger. The engineers insist that the roads with few accidents are safe, and that the complainers are neurotic.

Exposure

More sophisticated members of the orthodox school of risk assessment might raise questions about the appropriate measure of *exposure* of children to traffic. Although traffic has increased enormously since the 1920s, the exposure of children to traffic has probably decreased. It is not possible to measure this decrease as far back as the 1920s, but surveys conducted in England in 1971 and 1990 suggest a large reduction in children's exposure over that 19 year period in response to the perceived growth in the threat of traffic. In 1971, for example, 80 per cent of seven and eight year old children in England travelled to school on their own, unaccompanied by an adult. By 1990 this figure had dropped to 9 per cent; the questionnaire survey disclosed that the parents' main reason for not allowing their children to travel independently was fear of traffic.

But this evidence is still only broadly indicative of changes in exposure. It is far from providing a precise measure. Indeed, in the case of children's road safety, it is not clear how exposure might be measured. How might one measure the duration of their exposure to traffic, or the intensity of this exposure, or its quality? Children are impulsive, energetic and frequently disobedient (characteristics commonly found also in adults). They have short attention spans and their judgement in assessing traffic dangers varies greatly with age. They rarely walk purposefully to school or anywhere else. They frequently have balls or tin cans to kick about, or bicycles or skate boards on which to vie with one another in feats of daring. No survey evidence exists to chart changes over time in the time they spend playing *in* the street, as distinct from *alongside* the street, or the changes in speed, volume and variability of traffic to which they are exposed.

But the most intractable measurement problem of all is posed by changes in levels of *vigilance*, of both children and motorists, in response to variations in perceptions of the likelihood of colliding with one another. As perceived threats increase, people respond by being more careful. The variety of ways in which this might be done has defeated attempts to measure it. The unquantified, and mostly unquantifiable, changes in the exposure of children to traffic danger are characteristic of the difficulties encountered by *all* attempts to produce "objective" measures of risk outside of the casino where the odds can be mechanistically controlled.

The problem for those who seek to devise objective measures of risk is that people to varying degrees modify both their levels of vigilance and their exposure to danger in response to their *subjective* perceptions of risk. The modifications in response to an increase in traffic danger, for example, might include fitting better brakes, driving more slowly or carefully, wearing helmets or seat belts or conspicuous clothing, or withdrawing from the threat completely or – in the case of children no longer allowed to play in the street – being withdrawn from the threat by their parents.

In the physical sciences the test of a good theory is its ability to predict something that can be measured by independent observers. It is the Royal Society's purpose in studying risk – to manage it – that frustrates efforts to produce such predictions. Because both individuals and institutions respond to their perceptions of risk by seeking to manage it, they alter that which is predicted as it is predicted.

The league tables of "objective risk" commonly compiled to rank the probabilities of death from different causes – from "radiation" to "being a coal miner" – are constructed from data of immensely variable quality (see Ch. 5). But even if the accuracy and reliability of the data could be ensured, there would remain an insuperable problem in interpreting them as objective measures of risk for individuals. They are aggregate measures of past risk-taking by large and disparate populations of risk-takers, and they are a part of the evidence that shapes the perceptions of risk which influence future risk-taking.

All risks are conditional. It would, for example, be possible to devise a bowling machine that would, at randomized intervals, roll a child-size ball across a street. *If* the volume and speed of traffic were known, and *if* the speed and frequency of the balls were known, it would be possible to calculate the objective probability of ball and vehicle colliding within a specified time. But it would be useless as an estimate of the number of children one would expect to be knocked down on a residential street, because both children and drivers have expectations of each other's behaviour, observe it, and respond to it. The probabilities cited in the league tables would be valid predictors only *if* they could be concealed from the people affected, or *if* the people affected could be relied upon to behave as unresponsive dumb molecules; that is only *if* the known probabilities were not acted upon. And that, of course, would defeat the purpose of calculating them.

The response to risk: risk compensation

Figure 2.2, a model of the theory of risk compensation, illustrates the circularity of the relationships that frustrate the development of objective measures of risk. It is a model originally devised by Gerald Wilde in 1976, and modified by Adams (1985, 1988). The model postulates that
- everyone has a propensity to take risks
- this propensity varies from one individual to another

- this propensity is influenced by the potential rewards of risk-taking
- perceptions of risk are influenced by experience of accident losses – one's own and others'
- individual risk-taking decisions represent a balancing act in which perceptions of risk are weighed against propensity to take risk
- accident losses are, by definition, a consequence of taking risks; the more risks an individual takes, the greater, on average, will be both the rewards and losses he or she incurs.

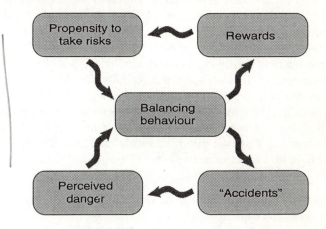

Figure 2.2   The risk "thermostat".

The arrows connecting the boxes in the diagram are drawn as wiggly lines to emphasize the modesty of the claims made for the model. It is an impressionistic, conceptual model, not an operational one. The contents of the boxes are not capable of objective measurement. The arrows indicate directions of influence and their sequence.

The balancing act described by this illustration is analogous to the behaviour of a thermostatically controlled system. The setting of the thermostat varies from one individual to another, from one group to another, from one culture to another. Some like it hot – a Hell's Angel or a Grand Prix racing driver for example; others like it cool – a Mr Milquetoast or a little old lady named Prudence. But no one wants absolute zero.

The young Roald Dahl, tricycling and whooping his way to school, exemplifies the need for excitement inherent in all of us. Psychologists sometimes refer to this as a need for a certain level of arousal. This need clearly varies, but there are no documented cases of its level being zero. And even if such a case could be found, it could not produce a zero-risk life. The single-minded pursuit of a zero-risk life by staying in bed would be likely, paradoxically, to lead to an early death from either starvation or atrophy. There is no convincing evidence that anyone wants a zero-risk life – it would be unutterably boring – and certainly no evidence that such a life is possible. The starting point of any theory of risk must be that everyone willingly takes risks. This is not the starting point of most of the literature on risk.

Homo prudens and free will

To err is human. So is to gamble. Human fallibility and the propensity to take risks are commonly asserted to be the root causes of accidents. How should responsibility for accidents be shared between these causes? The safety literature favours, overwhelmingly, human error as the cause of accidents (Reason 1990). No one wants an accident, therefore, it is argued, if one occurs it must be the result of a mistake, a miscalculation, a lapse of concentration, or simple ignorance of the facts about a dangerous situation.

Risk management in practice is overwhelmingly concerned not with balancing the costs and benefits of risk but with reducing it. Most of the literature on the subject is inhabited by *Homo prudens* – zero-risk man. He personifies prudence, rationality and responsibility. Large corporations such as Shell Oil hold him up as an example for all employees to emulate in their campaigns to eliminate all accidents.

> The safety challenge we all face can be very easily defined – to eliminate all accidents which cause death, injury and damage to the environment or property. Of course this is easy to state, but very difficult to achieve. Nevertheless, that does not mean that it should not be our aim, or that it is an impossible target to aim for. (Richard Charlton (1991), Director of Exploration and Production, Shell Oil)

> The aim of avoiding all accidents is far from being a public relations puff. It is the only responsible policy. Turning "gambling man" into "zero-risk man" (that is one who manages and controls risks) is just one of the challenges that has to be overcome along the way. (Koos Visser (1991), Head of Health, Safety and Environment, Shell Oil)

Homo prudens strives constantly, if not always efficaciously, to avoid accidents. Whenever he has an accident, it is a "mistake" or "error". When this happens, if he survives, he is acutely embarrassed and he tries, with the help of his expert advisers, to learn from his mistakes. Every major accident is followed by an inquiry into the events leading up to it in order to ensure that it can never happen again.

But people do willingly take risks. Gamblers may not like losing, but they do like gambling. Zero-risk man is a figment of the imagination of the safety profession. *Homo prudens* is but one aspect of the human character. *Homo aleatorius* – dice man, gambling man, risk-taking man – also lurks within every one of us. Perusal of films, television and the newspapers confirms that we live in a society that glorifies risk. The idols of the sports pages and financial pages are risk-takers. Our language is littered with aphorisms extolling the virtues of risk: "nothing ventured nothing gained", "no risk, no reward". Excessive caution is jeered at: "Prudence is a rich, ugly old maid courted by Incapacity" quoth William Blake.

The oil industry, which now seeks to turn its workers into zero-risk men, historically has been run by big risk-takers. The heroes of the industry in the public imagination are "wildcats" like Getty and the Hunts, or Red Adair – high-stakes gamblers prepared to put their lives or fortunes on the line. The workers involved in the processes of mineral exploration and production were "rough-necks" – men who were physically tough, and with a propensity to take risks. Their exploits have been enshrined in the folklore of the 49'ers of California, the Dangerous Dans of the Klondike, and innumerable other characters from mining areas all around the world. And the 49'ers have become a San Francisco football team, famous for its daring exploits.

We respond to the promptings of *Homo aleatorius* because we have no choice; life is uncertain. And we respond because we want to; too much certainty is boring, unrewarding and belittling. The safety literature largely ignores *Homo aleatorius*, or, where it does acknowledge his existence, seeks to reform him. It assumes that everyone's risk thermostat is set to zero, or should be.

Imagine for a moment that the safety campaigners were successful in removing all risk from our lives. What would a world without risk be like? A world without risk would be a world without uncertainty. Would we want it if we could have it? This is a question that has also been considered by some eminent scientists. Einstein argued about it with Max Born.

> You believe in a God who plays dice, and I in complete law and order in a world which objectively exists, and which I in a wildly speculative way, am trying to capture. I firmly *believe* [Einstein's emphasis], but I hope that someone will discover a more realistic way, or rather a more tangible basis than it has been my lot to find. Even the great initial success of the quantum theory does not make me believe in the fundamental dice game, although I am well aware that some of our younger colleagues interpret this as a consequence of senility. No doubt the day will come when we will see whose instinctive attitude was the correct one.
>
> Albert Einstein, in a letter to Max Born, 7 September, 1944 (Born 1971)

The uncertainty that quantum physicists *believe* is inherent in physical nature was anathema to Einstein. He *believed* that his inability to account for certain physical phenomena in terms of strict cause and effect was a consequence of his ignorance of the *laws* governing the behaviour of the phenomena in question, or his inability to measure them with sufficient precision.

Max Born, to whom he was writing, was one of the leading figures in the development of quantum mechanics. He was not persuaded by Einstein's strict determinism, but could not find evidence or argument to persuade

Einstein to accept the quantum theory. Einstein objected that "physics should represent reality in time and space free from spooky actions at a distance".

One of the reasons, perhaps the main reason, why Born preferred spooky actions at a distance to strict determinism was that "strict determinism seemed . . . to be irreconcilable with a belief in responsibility and ethical freedom". He lamented:

> I have never been able to understand Einstein in this matter. He was, after all, a highly ethical person, in spite of his theoretical conviction about predetermination . . . . Einstein's way of thinking in physics could not do without the "dice-playing God", to be absolutely correct. For in classical physics the initial conditions are not predetermined by natural laws, and in every prediction one has to assume that the initial conditions have been determined by measurement, or else to be content with a statement of probability. I basically believe that the first case is illusory
> . . .

And he wrote back to Einstein:

> Your philosophy somehow manages to harmonize the automata of lifeless objects with the existence of responsibility and conscience, something which I am unable to achieve.

Uncertainty, according to Born, is the only thing that permits us the possibility of moral significance. Only if there is uncertainty is there scope for responsibility and conscience. Without it we are mere predetermined automata.

In the realm of theology the debate about determinism goes back far beyond Born and Einstein. For centuries believers in free will have contended with believers in predestination. No resolution of the debate is in prospect. Theologians usually retire from the discussion with explanations like the following:

> The relation between the will of God and the will of man is mysterious. The former is eternal and irreversible, the latter real and free, within its proper limits. The appearance of contradiction in this arises from the finiteness of our understanding, and the necessity of contemplating the infinite and the immutable from a finite and mutable point of view. (Hall 1933)

In the face of uncertainty, both scientists and theologians fall back on belief. And even devout determinists such as Einstein cannot forgo the occasional use of the word "ought", which presupposes choice, and which brings in its train both uncertainty and moral significance. If the determinists are wrong, it would appear that risk is inescapable. It also appears to be desirable. Some risk-taking behaviour appears to be a confirmation of moral autonomy. Dostoevsky suggests that such confirmation in itself might be considered the ultimate reward for risk-taking. Only by invoking such a reward can one account for behaviour that would otherwise be seen as perverse and self-destructive. This is how Dostoevsky (1960) puts it:

> What man wants is simply *independent* choice, whatever that independence may cost and wherever it may lead . . . . Reason is an excellent thing, there's no disputing that, but reason is nothing but reason and satisfies only the rational side of man's nature, while will is a manifestation of the whole life . . . I repeat for the hundredth time, there is one case, one only, when man may consciously, purposely, desire what is injurious to himself, what is stupid, very stupid – simply in order to have the right to desire for himself even what is very stupid and not be bound by an obligation to desire only what is sensible. Of course, this very stupid thing, this caprice of ours, may be in reality, gentlemen, more advantageous for us than anything else on earth, especially in certain cases. And in particular it may be more advantageous than any advantage even when it does us obvious harm, and contradicts the soundest conclusions of our reason concerning our advantage – for in any circumstances it preserves for us what is most precious and most important – that is, our personality, our individuality.

From a Dostoevskian perspective, the greater the success of the safety regulators in removing uncertainty from our lives, the stronger will become the compulsion to reinstate it. This perspective has challenging implications for accident reduction programmes that attempt to promote safety by producing "failsafe" or "foolproof" environments, or by the use of rewards for safe behaviour, or penalties for dangerous behaviour. A world from which *all* risk had been removed would be one with no uncertainty or freedom or individuality. The closer one approaches such a state, the greater is likely to be the resistance to further progress, and the more likely will be outbreaks of Dostoevskian "irrationality".

Risk: an interactive phenomenon

Figure 2.2 above can be used to describe the behaviour of a motorist going around a bend in the road. His speed will be influenced by the rewards of risk. These might range from getting to the church on time, to impressing his friends with his skill or courage. His speed will also be influenced by his perception of the danger. His fears might range from death, through the cost of repairs following an accident, to embarrassment. It will also depend on his judgement about the road conditions: Is there ice or oil on the road? How sharp is the bend and how high the camber? – and the capability of his car – How good are the brakes, suspension, steering and tyres?

Overestimating the capability of the car, or the speed at which the bend can be safely negotiated, can lead to an accident. Underestimating these things will reduce the rewards at stake. The consequences, in either direction, can range from the trivial to the catastrophic. The safety literature almost universally ignores the potential loss of reward consequent on behaviour that is too safe. Its ambition is to eliminate *all* accidents.

Figure 2.3 introduces a second car to the road to make the point that risk is an interactive phenomenon. One person's balancing behaviour has consequences for others. On the road other motorists can impinge on your "rewards" by getting in your way and slowing you down, or help you by giving way. One is also concerned to avoid hitting them, or being hit by them. Driving in traffic involves monitoring the behaviour of other motorists, speculating about their intentions, and estimating the consequences of a misjudgement. If you see a car approaching at speed and wandering from one side of the road to another, you are likely to take evasive action, unless perhaps you place a very high value on your dignity and rights as a road user and fear a loss of esteem if you are seen to be giving way. During this interaction enormous amounts of information are processed. Moment by moment each motorist acts upon information received, thereby creating a new situation to which the other responds.

Figure 2.4 introduces another complication. On the road, and in life generally, risky interaction frequently takes place on terms of gross inequality. The damage that a heavy lorry can inflict on a cyclist or, pedestrian is great; the physical damage that they might inflict on the lorry is small. The lorry driver in this illustration can represent the controllers of large risks of all

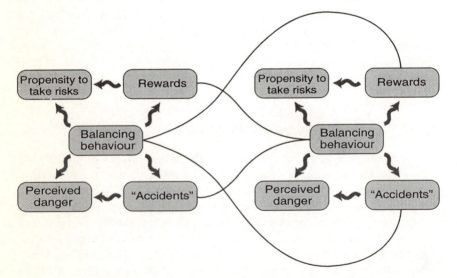

Figure 2.3   The risk "thermostat": two drivers interacting.

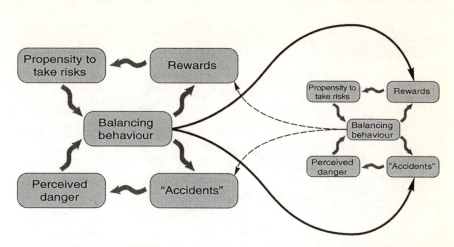

Figure 2.4   The risk "thermostat": lorry driver and cyclist interacting.

sorts. Those who make the decisions that determine the safety of consumer goods, working conditions or large construction projects are, like the lorry driver, usually personally well insulated from the consequences of their decisions. The consumers, workers or users of their constructions, like the cyclist, are in a position to suffer great harm, but not inflict it.

Problems of measurement

Risk comes in many forms. In addition to economic risks – such as those encountered in the insurance business – there are physical risks and social risks, and innumerable subdivisions of these categories: political risks, sexual risks, medical risks, career risks, artistic risks, military risks, motoring risks, legal risks . . . The list is as long as there are adjectives to apply to behaviour in the face of uncertainty. These risks can be combined or traded. In some occupations people are tempted by danger money. Some people, such as sky-diving librarians, may have very safe occupations and dangerous hobbies. Some young male motorists would appear to prefer to risk their lives rather than their peer-group reputations for courage.

Although the propensity to take risks is widely assumed to vary with circumstances and individuals, there is no way of testing this assumption by direct measurement. There is not even agreement about what units of measurement might be used. Usually the assumption is tested indirectly by reference to accident outcomes; on the basis of their accident records young men are judged to be risk seeking, and middle-aged women to be risk averse. But this test inevitably gets muddled up with tests of assumptions that accidents are caused by errors in risk perception, which also cannot be measured

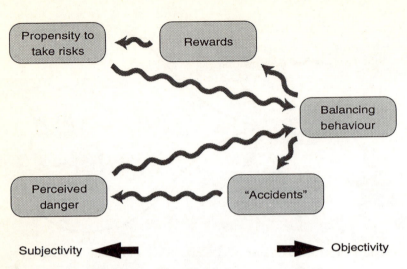

Figure 2.5   The risk "thermostat" stretched.

directly. If Nigel Mansell crashes at 180 mph in his racing car, it is impossible to determine "objectively" whether it was because he made a mistake or he was taking a risk.

Both the rewards of risk and the accident losses defy reduction to a common denominator. The rewards come in a variety of forms: money, power, glory, love, affection, (self)-respect, revenge, curiosity requited, or simply the sensation (pleasurable for some) accompanying a rush of adrenaline. Nor can accident losses be measured with a single metric. Road accidents, the best documented of all the realms of risk, can result in anything from a bent bumper to death; and there is no formula that can answer the question – how many bent bumpers equals one life?

Yet the Royal Society insists that there has to be such a formula. Its definition of detriment – "a numerical measure of the expected harm or loss associated with an adverse event" – requires that there be a scale "to facilitate meaningful additions over different events". Economists have sought in vain for such a measure. The search for a numerical measure to attach to the harm or loss associated with a particular adverse event encounters the problem that people vary enormously in the importance that they attach to similar events. Slipping and falling on the ice, as noted above, is a game for children, and an event with potentially fatal consequences for the elderly. The problems encountered in attempting to assign money values to losses will be discussed further in Chapter 6; the main problem is that the only person who can know the true value of an accident loss is the person suffering it, and, especially for large losses – such as loss of reputation, serious injury, or death – most people find it difficult, if not impossible, to name a cash sum that would compensate them for the loss.

Figure 2.5 is a distorted version of Figure 2.2 with some of the boxes displaced along an axis labelled "Subjectivity–Objectivity". The box which is displaced farthest in the direction of objectivity is "balancing behaviour". It is possible to measure behaviour directly. It was noted above, for example, that parents have withdrawn their children from the streets in response to their perception that the streets have become more dangerous. It is possible in principle to measure the decline in the amount of time that children spend in the streets exposed to traffic, but even here the interpretation of the evidence is contentious. Do children now spend less time on the street because they spend more time watching television, or do they spend more time watching television because they are not allowed to play in the streets? *All* of the elements of the risk compensation theory, and those of any contenders of which I am aware, fall a long way short of the objective end of the spectrum. Behaviour can be measured, but its causes can only be inferred.

And risks can be displaced. If motorcycling were to be banned in Britain, it would save about 500 lives a year. Or would it? If it could be assumed that all the banned motorcyclists would sit at home drinking tea, one could simply subtract motorcycle accident fatalities from the total annual road accident death toll. But at least some frustrated motorcyclists would buy old bangers and try to drive them in a way that pumped as much adrenaline as their motorcycling, and in a way likely to produce more kinetic energy to be dispersed if they crashed. The alternative risk-taking activities that they might get up to range from sky-diving to glue sniffing, and there is no set of statistics that could prove that the country had been made safer, or more dangerous, by the ban.

Figure 2.6, the dance of the risk thermostats, is an attempt to suggest a few of the multitudinous complications that flow from the simple relationships depicted in Figures 2.2 to 2.5. The world contains over 5 billion risk thermostats. Some are large; most are tiny. Governments and large businesses make decisions that affect millions if not billions. Individuals for the most part adapt as best they can to the consequences of these decisions. The damage that they individually can inflict in return, through the ballot box or market, is insignificant, although in aggregate, as we shall see in Chapter 3, they can become forces to be reckoned with.

Overhanging everything are the forces of nature – floods, earthquakes, hurricanes, plagues – that even governments cannot control, although they sometimes try to build defences against them. And fluttering about the dance floor are the Beijing Butterflies beloved of chaos theorists; they ensure that the best laid plans of mice, men and governments gang aft agley. Figure 2.6 shows but an infinitesimal fraction of the possible interactions between all the world's risk thermostats; there is not the remotest possibility of ever devising a model that could predict reliably all the consequences of intervention in this system. And chaos theorists now tell us that it is possible for very small changes in complex non-linear systems to produce very large effects.

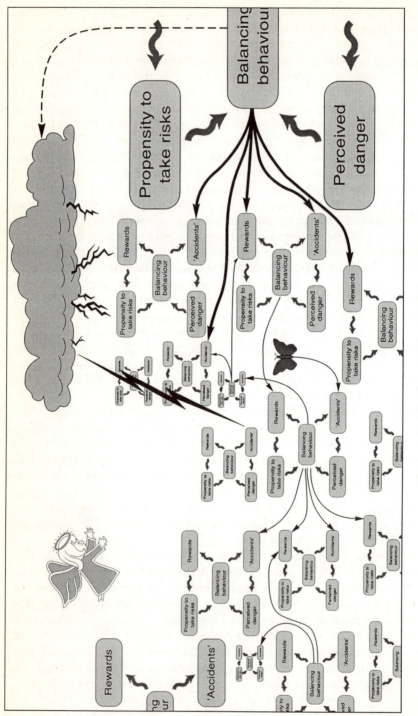

Figure 2.6 The dance of the risk thermostats.

And finally, Figure 2.6 includes a line (broken to indicate scientific uncertainty) to symbolize the impact of human behaviour on nature and all the natural forces that overhang the dance floor. Discussion of the second winged species, the angel, is reserved for Chapter 11.

The physical science involved in predicting such things as earthquakes and hurricanes is far from producing useful forecasts and even further from resolving the controversies over the greenhouse effect and ozone holes. Forecasting the weather more than a week ahead is still thought to be impossible, and when human beings get involved things become even more difficult. The clouds are indifferent to what the weather forecaster might say about them, but people respond to forecasts. If they like them, they try to assist them; if not they try to frustrate them.

The history of the diffusion of AIDS, for example, is very imperfectly known, not only because of problems associated with defining and measuring the symptoms of the disease, but also because it is a stigmatizing disease that people try to conceal, even after death; the order of magnitude of the number of people in the world now suffering from AIDS is still in dispute. The future course of the disease will depend on unpredictable responses to the perceived threat, by governments, by scientists and by individuals – and responses to the responses in a never-ending chain. Forecasts of the behaviour of the disease will inform the perception of the threat, and influence research budgets, the direction of research, and sexual practices, which will in turn influence each other.

Scientific uncertainty about the physical world, the phenomenon of risk compensation, and the interactive nature of risk all render individual events inherently uncertain.

Varieties of uncertainty

"Risk" and "uncertainty" have assumed the rôle of technical terms in the risk and safety literature since 1921, when Frank Knight pronounced in his classic work *Risk, uncertainty and profit* that
- if you don't know for sure what will happen, but you know the odds, that's *risk*, and
- if you don't even know the odds, that's *uncertainty*.

But in common, non-technical, usage this distinction between risk and uncertainty is frequently blurred and the words are used interchangeably. And even in the technical literature the distinction is often honoured in the breach. Virtually all the formal treatments of risk and uncertainty in game theory, operations research, economics or management science require that the odds be known, that numbers be attachable to the probabilities and magnitudes of possible outcomes. In practice, since such numbers are rarely available, they are usually assumed or invented, the alternative being to

25

admit that the formal treatments have nothing useful to say about the problem under discussion (see Ch. 6 on monetizing risk for examples).

The philosopher A. J. Ayer (1965), discussing the various senses in which the word *chance* is used to refer to judgements of probability made a useful threefold distinction. There are, he said, offering examples:

- *judgements of a priori probability*: "the chance of throwing double-six with a pair of true dice is one in 36",
- *estimates of actual frequency*: "there is a slightly better than even chance that any given unborn infant will be a boy", and
- *judgements of credibility*: "there is now very little chance that Britain will join the Common Market".

The first two of these senses, often combined in the form of inferential statistics, are the basis of most treatments of "objective risk". Allied to the law of large numbers, these treatments provide useful routine guidance in many areas of life. Insurance companies for example consult past claim experience in calculating the premiums they charge to cover future risks. But even insurance companies, with their actuaries, powerful computers and large databases, find this guidance far from infallible in the face of true uncertainty. At the time of writing, most insurance companies in Britain are reporting large losses, and some Lloyds Names are confronting bankruptcy.

Reports in the press of the plight of some of the Lloyds Names suggest that many of them did not appreciate the true nature of the business to which they had lent their names. To become a Lloyds Name, no investment is required. One must provide proof of personal wealth in excess of £250,000, and give an undertaking of *unlimited* liability for losses incurred by one's syndicates; a Name is then entitled to a share of the annual profits – the difference between the syndicates' premium income and payments of claims. It is now clear that many of the Names, although wealthy, were naïve. Lloyds' history of profitability had persuaded them that they had simply to lend their names and collect the profits; another case, they thought, of "to them that hath shall be given". But they had signed up for the uncertainty business.

Uncertainty as defined by Knight is inescapable. It is the realm not of calculation but of judgement. There are problems where the odds are known, or knowable with a bit more research, but these are trivial in comparison with the problems posed by uncertainty. Blake's picture of Newton concentrating on making precise measurements with a pair of callipers while surrounded by the mysterious "sea of time and space" (frontispiece) is an apt metaphor for the contribution that science is capable of making to the resolution of controversies rooted in uncertainty. Newton's approach can only ever deal with risk as narrowly defined by Knight and the Royal Society – as quantifiable probability. The concern of this book is with the more broadly defined risk of everyday English and everyday life – unquantifiable "*danger, hazard, exposure to mischance or peril*" (OED). Risk in these senses embodies the concepts of probability and magnitude found in the quantified "scientific" definitions of risk, but does not insist that they be precisely knowable. If one retreats from the unattainable aspiration of precise quantification, one may find, I believe, some useful aids for navigating the sea of uncertainty.

Chapter 3

PATTERNS IN UNCERTAINTY

The subjectivity and relativity of risk – so strongly resisted by those who aspire to a scientific, objective treatment of the subject – have respectable scientific antecedents. According to Einstein's theory of relativity, the size of an object in motion depends on the vantage point of the measurer. The length of a moving train carriage, to take a much-used illustration, will be longer if measured by someone travelling inside the carriage than if measured by someone standing beside the track. For the speeds at which trains travel, the difference will be negligible. The difference becomes highly significant as the speed of light is approached; at such speeds "objective" measures – independently verifiable measures – of speed, mass and time must be accompanied by a specification of the frame of reference within which they were made. And according to Heisenberg's uncertainty principle, the act of measuring the location of a particle alters the position of the particle in an unpredictable way.

Similar problems of relativity and indeterminacy confront those who seek to pin down risk with objective numbers. Risk is constantly in motion and it moves in response to attempts to measure it. The problems of measuring risk are akin to those of physical measurement in a world where everything is moving at the speed of light, where the act of measurement alters that which is being measured, and where there are as many frames of reference as there are observers.

A part of the dance of the risk thermostats described in Chapter 2 takes place in financial markets. Forecasters, market tipsters, even astrologers, predict the future course of currency exchange rates, share and commodity prices, interest rates, and gross domestic products. People buy and sell, guided by their expectations, which are modified in the light of the behaviour of other buyers and sellers. Other dancers can be found in supermarkets pondering information about calories and cholesterol and fibre and pesticides, while the supermarket owners monitor their concerns and buying behaviour through questionnaires and, sometimes, closed circuit television – customer and owner constantly modify their behaviour in response to that of the other. Dancers can also be found in ministries of defence all around the world, spying on each other, and spending vast sums of money to defend themselves against the "defence" of the enemy. And still others can be found on the road. Throughout the world hundreds of millions of motor vehicles mix with billions of people. A pedestrian crossing a busy road tries to make eye contact with an approaching motorist. Will he slow down? The motorist tries to divine the intentions of the pedestrian. Will he give way? Shall I? Shan't I? Will he? Won't he? *As the distance between them closes, signals, implicit and explicit, pass between them at the speed of light. Risk perceived is risk acted upon. It changes in the twinkling of an eye as the eye lights upon it.*

"Risk" is defined, by most of those who seek to measure it, as the product of the probability and utility of some *future* event. The future is uncertain and inescapably subjective; it does not exist except in the minds of people attempting to anticipate it. Our anticipations are formed by projecting past experience into the future. Our behaviour is guided by our anticipations. If we anticipate harm, we take avoiding action. Accident rates therefore cannot serve, even retrospectively, as measures of risk; if they are low, it does not necessarily indicate that the risk was low. It could be that a high risk was perceived and avoided.

It is the very determination of the measurers to *change* risk that frustrates their ability to measure it. In carefully contrived circumstances – such as the spinning of a roulette wheel, or the example offered in Chapter 2 of the child-size ball rolled across a road – one can estimate objective probabilities for specified events. But in the real interactive world of risk management, where the purpose of measurement or estimation is to provide information to guide behaviour, risk will be contingent on behavioural responses, which will be contingent on perceived risk, and so on. And even where the probability of the event itself may lie beyond the control of the measurer – as in the case of a predicted meteor impact – the information will still have consequences. Some might pray, others might get drunk, others might dig deep burrows in which to shelter; and, if the event did not happen, it would be likely to take some time for society to recover and get back to business as usual. If the event were an astronomical certainty, it would, of course, not be a risk.

In Chapter 2 it was noted that the direction of the change intended by the risk measurers and managers is almost invariably down. The purpose of risk research is risk reduction. Funds are made available for such research in the hope or expectation that it will lead to a lowering of risk. The ultimate objective is often proclaimed to be the removal of *all* risk, the elimination of *all* accidents. Governments around the world continue to add to the existing libraries full of safety laws and regulations. Safety campaigners are relentless in their efforts to make the world safer. On achieving a regulatory or statutory goal, they do not stop. They identify a new risk, and press for new laws or regulations, or stricter enforcement of existing ones.

The world's largest industry

The relentless pursuit of risk reduction has made safety an enormous indus-try. It is not possible to be precise about its size, because safety merges with everything else – manufacturers of safety glass, for example, produce both glass and a safety product. But a few figures from some of the areas in which it operates in Britain will be sufficient to demonstrate the scope and eco-nomic significance of the risk reduction industry:

- *Safety in the home*. Thousands are employed devising and enforcing the regulations specifying the methods and materials used in house con-struction. Thousands more are employed in the caring professions to protect the elderly by installing hand-rails and skid-proof bath mats, and overseeing the welfare of children.
- *Fire*. There are 40,000 firemen in Britain, plus administrative support staff, fire engineers, fire consultants, fire insurers, fire protection serv-ices, people who make smoke alarms, people who make fire doors, peo-ple who install them, and a fire research station to help devise further protection.
- *Casualty services*. In addition to those employed in hospital casualty departments there are 20,000 ambulance drivers in the country and a further 70,000 St John's ambulance volunteers trained in first aid.
- *Safety at play*. Thousands more are employed to ensure safe play by inspecting playgrounds, installing rubberized surfacing, working as play supervisors and lifeguards in swimming pools, plus the people who train all these people.
- *Safety at work*. The Health and Safety Executive responsible for over-seeing safety at work employs 4,500 people, but this is the tip of the ice-berg. At University College London, where this book is being written, we have six full-time staff devoted to supervising our safety, plus a 13-person central safety committee, plus 97 members of staff desig-nated as safety officers, plus 130 who have qualified in first aid by tak-ing a one-week course sponsored by the college. I am writing behind a fire-resistant door, in a building made out of glass and concrete and steel, with frequent fire drills, and windows that will not open wide enough for me to fall out – or jump out. In factories safety is taken more seriously; goggles, helmets and steel-capped boots must be worn, machinery guarded, and "failsafe" procedures observed.
- *Safety on the road*. The most highly regulated activity of all is motoring. Most road traffic law is justified on safety grounds. Annually the number of people proceeded against in court for motor vehicle offences is over 2.5 million. They account for 75 per cent of all court proceedings and an enormous amount of police time. In addition, motorists pay fines for over 5 million fixed-penalty offences (mostly parking offences), but many of these are proceeded against for safety reasons.

Involved in all this legal work are 27,000 magistrates and judges, and untold numbers of police, lawyers, administrators and statisticians. Over 0.5 million motorists are tested for alcohol every year, almost 2 million new drivers are tested, and 19 million motor vehicles are given safety checks. Vehicle safety regulations cover almost every aspect of motor vehicles – the tread of tyres, the type of glass in windscreens, brakes, crash-worthiness – and add billions of pounds to the total annual cost. Safety also forms a major part of the justification for most new road building, an activity valued at over £2 billion a year, and employing many thousands.

Beyond all this there are the police forces employing over 150,000 people, the security industry selling personal and property protection, and reputed to be the world's fastest growing industry, the insurance industry, environmental health officers and pollution inspectors, the safety and environmental pressure groups, and the National Health Service, employing 1.5 million. Affecting the lives of even more people are measures addressing mega-risks such as nuclear power, ozone holes and the greenhouse effect. And the armed forces employ a further 300,000 and command an annual budget of £19 billion, to defend Britain against the risk of attack from other countries who spend similar or larger amounts of money on "defence". And finally, world wide, there are the billions of part-time self-employed – all of us – routinely monitoring our environment for signs of danger and responding as we judge appropriate. Definitional problems preclude precision, but, when all its constituent parts are combined, the risk reduction industry almost certainly deserves to be called the world's largest industry.

Is the world getting value for money from the vast resources committed to risk reduction? A clear answer is seldom possible. Is, for example, the relatively small number of fatalities thus far attributable to the nuclear industry proof of safety-money effectively spent, or is it proof of money wasted on unnecessary defence against an exaggerated threat? It is impossible to say; no one can prove what would have happened had cheaper, less stringent, design standards and safety procedures been adopted. No one can foretell the frequency of future Chernobyls and, given the long latency period of low-dose radiation before health effects become apparent, no one can say what the ultimate cost of Chernobyl will be. And, the proponents of nuclear power might ask, if this cost could be known, might it be judged a price worth paying for the benefits of nuclear power?

The world's largest industry in all its manifestations from the provision of skid-proof bath mats to the design of nuclear containment vessels and Star Wars defence shields, appears to be guided by nothing more than hunch and prejudice – billions of hunches and billions of prejudices. The dance of the risk thermostats appears, at first sight, to be an inchoate, relativistic shambles. But further scrutiny discloses order and pattern in the behaviour of the dancers.

Patterns in uncertainty

The frustration of scientists attempting to measure risk suggests the direction in which we ought to turn for help. Ever since Thomas Kuhn's (1962) *The structure of scientific revolutions*, scientific frustration has been seen as symptomatic of paradigm conflict – of discord between the empirical evidence and the expectations generated by the scientists' paradigm, or world view. The history of science as described by Kuhn is a process of paradigm formation and paradigm overthrow in a never-ending quest for truth and meaning. The actors in this drama are characterized by their constellations of assumptions and beliefs about the nature of the reality they are exploring. Occasionally some of these beliefs assume the form of explicit hypotheses that can be tested, but mostly they are implicit and subconscious. Risk, however, presents problems and challenges not just to scientists but to lay persons as well. We all, daily, have our world views confronted by empirical evidence. And the world about which we must take a view when we are making decisions about risk comprises not just physical nature, but human nature as well.

Douglas & Wildavsky (1983) began their book *Risk and culture* with a question and an answer: "Can we know the risks we face now and in the future? No, we cannot; but yes, we must act as if we do." The subtitle of their book was *An essay on the selection of technological and environmental dangers*, and the question they were addressing was why some cultures face risk *as if* the world were one way, and others *as if* it were very different. Why do some cultures select some dangers to worry about where other cultures see no cause for concern?

The management of ecosystems such as forests, fisheries or grasslands provides a good example of the practical consequences of behaving as if the world were one way rather than another. Ecosystem managers must make decisions in the face of great uncertainty. Ecologists who have studied managed ecosystems have found, time and again, that different managing institutions faced with apparently similar situations have adopted very different management strategies. Holling (1979, 1986) discerned patterned consistencies in these differences that appeared to be explicable in terms of the managers' *beliefs* about nature. He noted that, when confronted by the need to make decisions with insufficient information, they *assumed* that nature behaves in certain ways. He reduced the various sets of assumptions he encountered to three "myths of nature" – nature benign, nature ephemeral, and nature perverse/tolerant. Schwarz & Thompson (1990) added a fourth – nature capricious – to produce the typology illustrated by Figure 3.1. The essence of each of the four myths is illustrated by the behaviour of a ball in a landscape, and each, they concluded was associated with a distinctive management style.

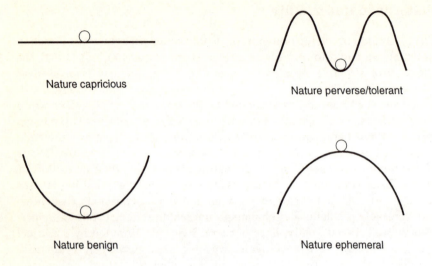

Figure 3.1 The four myths of nature.

- *Nature benign:* nature, according to this myth, is predictable, bountiful, robust, stable, and forgiving of any insults humankind might inflict upon it; however violently it might be shaken, the ball comes safely to rest in the bottom of the basin. Nature is the benign context of human activity, not something that needs to be managed. The management style associated with this myth is therefore relaxed, non-interventionist, *laissez-faire.*
- *Nature ephemeral:* here nature is fragile, precarious and unforgiving. It is in danger of being provoked by human carelessness into catastrophic collapse. The objective of environmental management is the protection of nature from humans. People, the myth insists, must tread lightly on the Earth. The guiding management rule is *the precautionary principle.*
- *Nature perverse/tolerant:* this is a combination of modified versions of the first two myths. Within limits, nature can be relied upon to behave predictably. It is forgiving of modest shocks to the system, but care must be taken not to knock the ball over the rim. Regulation is required to prevent major excesses, while leaving the system to look after itself in minor matters. This is the ecologist's equivalent of a mixed-economy model. The manager's style is *interventionist.*
- *Nature capricious:* nature is unpredictable. The appropriate management strategy is again *laissez-faire,* in the sense that there is no point to management. Where adherents to the myth of nature benign trust nature to be kind and generous, the believer in nature capricious is agnostic; the future may turn out well or badly, but in any event it is beyond his control. The non-manager's motto is *que sera sera.*

Myths of human nature

The four myths of nature are all anthropocentric; they represent beliefs not just about nature but about humankind's place *in* nature. The four myths of nature, by focusing attention on the managers' *beliefs*, have proved remarkably fruitful in helping to make sense of the things that managers do. They carry out their responsibilities *as if* nature could be relied upon to behave in a particular way.

The central theme of *Risk and culture* is that risk is "culturally constructed". This theme has been further refined, and linked to Holling's myths of nature, by Schwarz & Thompson (1990) in *Divided we stand* and by Thompson et al. (1992) in *Cultural theory*. In these works the authors inquire into the origins of the beliefs about nature that guide risk-taking decisions and, like Holling, they discern patterns. The essence of these cultural patterns has also been distilled into a fourfold typology, illustrated by Figure 3.2.

This typology, originally known rather cryptically as "grid/group", has two axes. Moving along the horizontal axis from left to right, human nature becomes less individualistic and more collectivist. The vertical axis is labelled "prescribed/unequal" and "prescribing/equal"; at the top, human

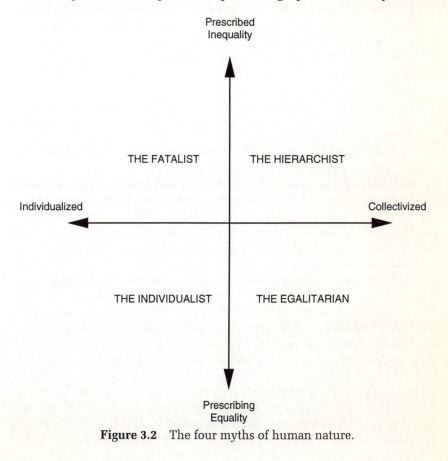

Figure 3.2 The four myths of human nature.

behaviour is "prescribed" – constrained by restrictions on choice imposed by superior authority, and social and economic transactions are characterized by inequality. At the bottom there are no externally prescribed constraints on choice; people negotiate the rules as equals as they go along.

In the lower left-hand corner of this diagram we find the *individualist* and in the upper right-hand corner the *hierarchist*. These are familiar characters to sociologists and anthropologists accustomed to the division of cultures into those organized as hierarchies and those in which markets mediate social and economic relations. This traditional bi-polar typology has been expanded by the cultural theorists to include two new archetypes, the *egalitarian* and the *fatalist*. A full description of this typology can be found in *Divided we stand* and *Cultural theory*; in brief:

- *Individualists* are enterprising "self-made" people, relatively free from control by others, and who strive to exert control over their environment and the people in it. Their success is often measured by their wealth and the number of followers they can command. The self-made Victorian mill owner would make a good representative of this category.
- *Hierarchists* inhabit a world with strong group boundaries and binding prescriptions. Social relationships in this world are hierarchical, with everyone knowing his or her place. Members of caste-bound Hindu society, soldiers of all ranks, and civil servants, are exemplars of this category.
- *Egalitarians* have strong group loyalties but little respect for externally imposed rules, other than those imposed by nature. Group decisions are arrived at democratically and leaders rule by force of personality and persuasion. Members of religious sects, communards, and environmental pressure groups all belong to this category.
- *Fatalists* have minimal control over their own lives. They belong to no groups responsible for the decisions that rule their lives. They are non-unionized employees, outcasts, untouchables. They are resigned to their fate and they see no point in attempting to change it.

In *Divided we stand* Schwarz & Thompson proposed that this typology of *human nature* could be mapped onto the typology of *physical nature*. Figure 3.3 illustrates this mapping. World views they argued were inseparable from *ways of life*, and *viable* ways of life were those with world views that helped them to survive in the face of the uncertainty of physical nature, and also in the face of competing world views. The capriciousness of nature, they suggest, complements, and is complemented by, a sense of fatalism. A capricious nature cannot be governed; one can only hope for the best, and duck if you see something about to hit you. Individualism accords with a benign nature that provides a supportive context for the individualist's entrepreneurial, trial-and-error way of life. An ephemeral nature demands that we tread lightly on the Earth and complements the "small-is-beautiful" ethic of the egalitarian. And the perverse/tolerant view of nature complements the

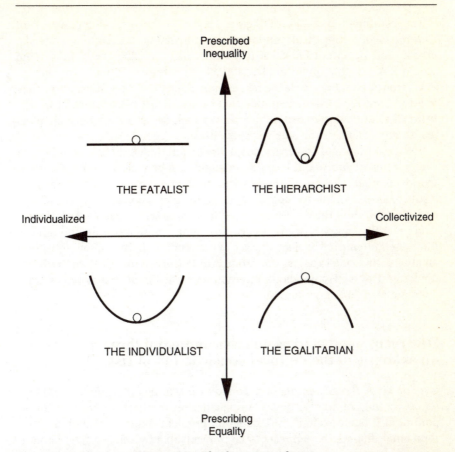

Figure 3.3 The four rationalities.

hierarchist's managerial approach to both nature and social relations; research is needed to identify the limits of nature's tolerance, and regulation is required to ensure that the limits are not exceeded.

These four distinctive world views are the basis of four different *rationalities*. Rational discourse is usually recognized by its adherence to the basic rules of grammar, logic and arithmetic. But in an uncertain world the premises upon which rational arguments are constructed are themselves beyond the reach of rationality. Disputes about risk in which the participants hurl charges of stupidity and irrationality at each other are usually seen upon dispassionate inspection to be arguments in which the participants are arguing from different premises, different paradigms, different world views – different myths of nature, both physical and human. These different rationalities tend to entrench themselves. Both the paradigms of science and the myths of cultural theory are powerful filters through which the world is perceived, and they are reinforced by the company one keeps.

The combined typology of Figure 3.3 forms the central framework of *Cutural theory*. Empirical support for the theory is, the authors concede, sparse, and in its current form it presents many challenges to those who would frame it as a quantifiable, refutable hypothesis. "What", the authors ask, "would count as evidence against our theory?" "Most damaging", they answer, "would be a demonstration that values are little constrained by institutional relationships." But values, as we shall see in subsequent chapters (especially Ch. 6), are as elusive as risk itself.

Both scientists and "ordinary people" confront the world armed only with their myths of nature. Cultural theory might best be viewed in the uncertain world we inhabit as the anthropologists' myth of myths. The validity of such a super-myth is not to be judged by the statistician's correlation coefficients and *t*-tests, but by the degree to which it accords with people's experience. And its utility can be judged only by the extent to which people find it helpful in their attempt to navigate the sea of uncertainty. In the following section an attempt is made to assess the validity and utility of this myth of myths by applying it to a dispute about an environmental threat that is typical of a wide range of such arguments.

Divergent perspectives on environmental threats: an example of the cultural construction of risk

Capital Killer II: still fuming over London's traffic pollution (Bell 1993) is a report by the London Boroughs Association on the health effects of traffic pollution. It is an example of a common problem – the never-ending environmental dispute that appears to be unresolvable by science. The London Boroughs Association (LBA) has a membership of 19, mainly Conservative-controlled London boroughs plus the City of London. As the subtitle of this report indicates, the association is unhappy – indeed fuming – over the lack of action by the Conservative central government to reduce traffic pollution in London. The report complains about the lack of resources provided by central government to deal with "this most serious of issues", and reproaches the government for its failure to follow its own policy advice as propounded in its White Paper on the Environment: "Where there are significant risks of damage to the environment, the government will be prepared to take precautionary action . . . even where scientific knowledge is not conclusive". The particular precautionary action that the LBA seeks is "action . . . now to reduce levels of traffic and pollution in London".

Why should political allies (or at least politicians belonging to the same party) fall out over this issue? Why should the local government politicians see an urgent need for action where their central government counterparts plead that there is insufficient evidence to justify such action? Let us look at the evidence on the health effects of traffic pollution summarized in the LBA report. The summary begins by citing the conclusions of its 1990 report *Capital Killer*: "exhaust emissions from road vehicles *may* cause major health problems. Since publication of the report research has continued *to suggest* links between air pollution and health." It accepts that "it will be *difficult to get hard information* on the long-term effects of air pollution on health". It says that "the link between air pollution and health is *not proven* but research is increasingly *suggesting* that there is such a link".

The report notes an alarming fivefold increase in the number of hospital admissions nationally for childhood asthma between 1979 and 1989 and says "*it may well be*" that air pollution is one of the factors contributing to the increased incidence and severity of asthma, and that "traffic exhausts *may* exacerbate" asthma and allergic disease. Unfortunately for this hypothesis, the magnitude of the changes in traffic and emissions between 1979 and 1989 are small relative to the health effects they are invoked to account for. Although childhood asthma is reported to have increased by 400 per cent, traffic nationally increased by only 57 per cent, and in urban areas, where the concentrations of pollutants are greatest, by only 27 per cent; further, during this period, reported emissions of nitrogen oxides, sulphur dioxide and lead decreased, and emissions of carbon monoxide increased by only 8 per cent and volatile organic compounds by only 3 per cent (DOT 1990). For such small changes in traffic and emissions to account for such a large change in the incidence of asthma requires a sensitivity of response for which the report presents no evidence.

The report goes on to look at the evidence for a link between traffic pollution and hay fever. It reports a theory from one study that car fumes damage the lining of the nose "and *could explain* why hay fever in cities is now three times more prevalent than in rural areas"; it cites another report that found that "more people *appear* to be suffering from hay fever", and yet another that creates "*suspicion* that worsening pollution is responsible" for the increased incidence of hay fever.

The report presents more evidence in similar, heavily qualified, vein, and then quotes, in tones of disappointment and incredulity, the government's response to the evidence: "there is [the Government says] *perceived to be* a growth in the incidence of respiratory illnesses, and many respiratory physicians do *believe* that there is an increase in the prevalence of asthma; but suggestions that the change in asthma levels is as a result of air pollution remain unproven". In the previous two paragraphs all the italics have been added by me to stress the LBA's acknowledgement of the tenuous nature of the evidence linking air pollution and ill health. In this paragraph the italics have been added by the author of the LBA report. The LBA's italics appear to be intended to encourage a sense of anger and incredulity in the reader. How, the report seems to ask, could the Government respond to such compelling evidence by suggesting that it was mere perception and belief? How could the Government not be moved to action?

The four rationalities as contending paradigms

These contrasting responses to the same evidence, or lack of it, provide an excellent example of *the cultural construction of risk*. This phenomenon can be found at work wherever disputes about health and safety are unresolved, or unresolvable, by science. For years the nuclear industry and environmentalists have argued inconclusively, producing mountains of "scientific" evidence in the process. Food safety regulations, AIDS, the greenhouse effect, seat belts and bicycle helmets are but a few other examples, from many that could be chosen, of health and safety controversies that have generated a substantial "scientific" literature without generating a consensus about what should be done. In all these cases, and a multitude of others, the participants commonly cast aspersions on the rationality of those who disagree with them. The approach of cultural theory suggests not that some are rational and others irrational, but that the participants are arguing rationally from different premises. This can be illustrated by the disagreement between the LBA and the Government about traffic pollution in London.

Individualists tend to view nature as stable, robust and benign, capable of shrugging off the insults of man, and rarely retaliating. They are believers in market forces and individual responsibility, and are hostile to the regulators of the "nanny State". For them the use of seat belts and helmets and the risks associated with sexual behaviour should be matters of individual discretion. The safety of food, like its taste and appearance, they would leave to the market. Where evidence is inconclusive, as in the case of the greenhouse effect and the health effects of air pollution, they place the onus of proof on those who would interfere. They tend to an optimistic interpretation of history, and are fond of citing evidence of progress in the form of statistics of rising gross domestic product and lengthening life expectancy.

Egalitarians cling to the view of nature as fragile and precarious. They would have everyone tread lightly on the Earth and in cases of scientific doubt invoke the precautionary principle. They join the individualists in opposition to the compulsory use of bicycle helmets and seat belts, but for different reasons; they argue that compelling people to wear helmets inhibits the use of an environmentally benign form of transport, and that seat belts and other measures that protect people in vehicles encourage heedless driving that puts pedestrians and cyclists at greater risk. AIDS confirms their view of the need for prudent and cautious behaviour in a dangerous world. The precariousness of individual health justifies protective measures in the form of food safety regulations. The greenhouse effect and the health effects of traffic pollution are both issues that cry out for the application of the precautionary principle. Egalitarians incline to an anxious interpretation of history; they read it as a series of dire warnings and precautionary tales of wars, plagues and famines, and species and civilizations extinguished through human greed or carelessness.

Hierarchists believe that nature will be good to them, *if* properly managed. They are members of big business, big government, big bureaucracy. They are respecters of authority, both scientific and administrative; those at the top demand respect and obedience, those at the bottom give it, and those in between do some of each. They believe in research to establish "the facts" about both human and physical nature, and in regulation for the collective good. If cyclists or motorists do not have the good sense to wear helmets or belt up, they should be compelled to do so. Food safety regulation, accompanied by monitors and enforcers to ensure compliance, is required to protect us from the careless or unscrupulous. Pending the discovery by scientists of an AIDS vaccine, sex is an activity demanding education, moral instruction or condoms, depending on the hierarchy's particular religious or secular persuasion. Hierarchists take a "balanced" view of history; it contains warnings but also holds out the promise of rewards for "correct" behaviour.

Fatalists, the fourth of cultural theory's categories, believe nature to be capricious and unpredictable. They hope for the best and fear the worst; life is a lottery over whose outcome they have no control. They tend to be found at the bottom of the socioeconomic heap, and hence are exposed to more than their share of traffic pollution, but they do not get involved in arguments about what should be done about it because they see no point; nothing they do will make any difference. They have high death rates both from "natural causes" and accidents. They do not study history.

These representatives of the categories of cultural theory are caricatures. Real people are more complex. But it is nevertheless possible in an examination of most long-running debates about health and safety to identify approximations of such caricatures among the leading participants.

The cultural construction of pollution

The debate about the health effects of traffic pollution is unlikely to be settled conclusively for some time, if ever. Describing the risks of traffic pollution as culturally constructed is not to say that they are mere figments of fevered imaginations. There is an obvious cause for concern; the exhaust emitted by cars contains many toxins, plus carbon dioxide, which is implicated in the greenhouse effect – a scientific controversy in its own right (see Ch. 9). The toxins are dispersed unevenly, in highly diluted form, over long periods of time. Some may be concentrated in the food chain, others may be transported great distances and/or combined with other pollutants to be rendered harmless, or more damaging. The environment into which they are dispersed contains plants, animals and people with varying susceptibilities to different toxins. Some toxins will be persistent and their effects cumulative. Some might have direct effects, others might weaken immune systems with results being manifest in the symptoms of a wide variety of opportunistic diseases. There are often long time-lags between exposure to pollutants and consequent damage to health, and most of the symptoms of illness caused by the ingredients of exhaust emission could have other causes. Some emissions might even be beneficial; in certain circumstances, for example, acid rain and carbon dioxide increase plant yields.

With few exceptions the toxic nature of the emissions is not in dispute. The unresolved question is whether they are emitted in quantities that cause significant damage, and if so whether this damage outweighs the benefits of the activity, motoring, that produces them. The LBA report reviewed here ostensibly addresses the health effects of traffic emissions, but it does so in the context of a wider debate about the benefits of a Government transport policy that is fostering an increase in traffic. Both the Government and the LBA are agreed that the evidence linking pollution to ill health is somewhat tenuous. They disagree about the appropriate policy response. This suggests that the real difference between them lies not in their view of the damage done by traffic emissions, but in their view of the benefits of the traffic. If the benefits are considered great, then the evidence required to justify a sacrifice of some of these benefits to reduce emissions should be compelling. The smaller the benefits, the stronger becomes the case for invoking the precautionary principle. And if the benefits are considered *negative* then even a suspicion that damage might result from emissions becomes an *additional* justification for curbing traffic.

Adding cultural filters to the risk thermostat

Both the *perceived danger* of pollution from traffic and the perception of the *rewards* of growing traffic will influence the *balancing behaviour* described by Figure 2.3: anxious cycling environmentalists wear masks when cycling in traffic, and campaign for measures to reduce traffic pollution; the Mr Toads in their large cars remain cheerfully oblivious; the Government seeks to "balance" "legitimate concerns about the environment" and "legitimate aspirations of motorists"; and the fatalist continues to stand at the bus stop inhaling traffic fumes while waiting for the bus that never comes. These diverse behavioural responses to the same objective reality imply that reality is filtered by paradigms, or myths of nature, both physical and human. Figure 3.4, in which the risk thermostat is fitted with cultural filters, suggests a way of combining the phenomena of risk compensation with the insights of cultural theory.

These filters are resistant to change, but they are not immutable. The positions adopted in the LBA report and the Government's response to it reflect a variety of pressures. They exhibit the biases of the civil servants responsible for writing the report and drafting the response. They are influenced by the views of the politicians responsible for commissioning the report, and the

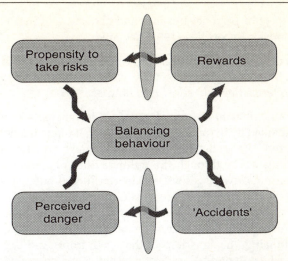

Figure 3.4   The risk thermostat with cultural filters.

pressures to which they are subjected by their constituents, who in turn are influenced by all the multifarious forces at work shaping public opinion. In a dispute such as that over traffic emissions and what should be done about them, explanations for the longevity of the argument and the inability of scientific evidence to resolve it are more likely to be found not in more scientific evidence, but in an examination of the sources of bias in the participants. Why should the LBA see traffic emissions as an urgent problem, while the Government dismisses their case as unproven?

One clue is provided by the LBA report. It advocates a few measures to monitor air pollution, and to encourage further research on the effects of air pollution. But most of its recommendations assume that the case against emissions is proven; the main thrust of the recommendations is that traffic should be reduced. It recommends that the Government and local authorities should invest more in the railways, increase taxes on car use, end subsidies for company cars and off-street parking, lower speed limits and increase their enforcement, calm traffic, provide their employees with bicycles and implement a 1,000-mile cycle route network in London, discourage out-of-town development, develop light rail, provide priority routes for buses, make better provision for people with disabilities, and provide secure parking for bicycles at stations.

The fact that this impressive list of recommendations emerges from a survey of the effects of traffic emissions on health, in which almost all the evidence was characterized in one way or another as inconclusive, suggests that the LBA and the Government have very different views about the desirability of the activity generating the emissions. The evidence for such a difference is compelling. While the LBA seeks to reduce dependence on the car, the Government "welcomes the continuing widening of car ownership as an

43

important aspect of personal freedom and choice", and in a recent speech Transport Minister Robert Key (1993) declared

> I love cars of all shapes and sizes. Cars are a good thing . . . . I also love roads . . . . The car is going to be with us for a long time. We must start thinking in terms that will allow it to flourish.

The car is an *individualist* form of transport. It transports its owner, sometimes accompanied by invited guests, in privacy. It offers freedom, and independence from the constraints of public transport. Its benefits and advantages pass easily through the cultural filter of the individualist, its disbenefits and disadvantages are efficiently filtered out, especially in the suburbs and the countryside where most Conservative voters live and where there is still driving room. But in urban areas the individualist increasingly experiences cognitive dissonance. The car provides the freedom to sit in traffic jams inhaling the emissions of the car in front. And traffic jams provide time to reflect upon the merits of alternative transport policies.

Most of the older built environments in Britain were built to a scale that cannot allow the car to flourish. These environments were designed for pedestrians and cyclists and, for journeys beyond the range of these modes of travel, users of public transport. Attempts to accommodate the still-growing numbers of cars are causing damage that is perceptible to even the most resolute individualist. Market forces do not appear to be providing solutions. There is no space for more roads. The need for restraint and some alternative way of getting about becomes increasingly difficult to resist.

The *hierarchists* of the Department of Transport and the motor industry offer their services. Transport is in a mess because it is badly or insufficiently regulated. They insist on catalytic converters and lead-free petrol to reduce emissions. They work on cleaner, more efficient engines, and traffic control systems to make more efficient use of the existing road system. They commission research into electronic road pricing, computer-based route-guidance systems, and "intelligent" cars and roads. The problem of road safety demands even more engineering and regulation, and education to foster responsible attitudes. The use of seat belts and helmets is made compulsory. Road safety education inculcates attitudes of deference to the car. Barriers are erected to force pedestrians over footbridges or through tunnels. Ever more safety is "engineered" into cars. New roads are built to separate people and traffic. It is all justified by cost–benefit analysis.

The *egalitarians* also enter the fray. For the Friends of the Earth and other environmentalists, the damage done by the car is symptomatic of a deeper malaise. Runaway economic growth, unbridled materialism, and the hubris of science and technology threaten global catastrophe. The egalitarian filter blocks many of the benefits of growth, materialism, science and technology, while allowing through and magnifying all evidence of the destructiveness of these forces, and even threats of destructiveness. Their solutions to our transport problems focus on modes of travel that are environmentally benign and socially constructive. The car does violence to their communal ethos; walking, cycling and public transport promote community spirit.

The *fatalists* have no comment to offer. They do not participate in policy debates.

Groping in the dark

The above speculations are relevant to *all* disputes that are unresolved or unresolvable by science. Wherever the evidence is inconclusive, the scientific vacuum is filled by the assertion of contradictory certitudes. For the foreseeable future scientific certainty is likely to be a rare commodity, and issues of health and safety – matters of life and death – will continue to be decided on the basis of scientific knowledge that is not conclusive. The conventional response to this unsatisfactory state of affairs is to assert the need for more science.

More trustworthy scientific information will do no harm, but the prospect is remote of settling most current controversies within the time available to make decisions; where adherents to the precautionary principle perceive the possibility of serious harm, they press for action as a matter of urgency. Just how remote the prospect of scientific resolution, and how large the scientific vacuum, can be illustrated graphically with the help of some numbers taken from the 1983 report by the National Research Council in the USA entitled *Risk assessment in the Federal Government: managing the process*. This report was the product of a study whose purpose was "to strengthen the reliability and the objectivity of scientific assessment that forms the basis for federal regulatory policies applicable to carcinogens and other public health hazards". The report noted that about 5 million different chemical substances are known to exist, and that their safety is theoretically under regulatory jurisdiction. Of these, it pointed out, fewer than 30 have been definitely linked to cancer in humans, 1,500 have been found to be carcinogenic in tests on animals, and about 7,000 have been tested for carcinogenicity.

The black rectangle in Figure 3.5 represents the darkness of ignorance: what we do *not* know about the carcinogenic effects of most substances. The size of the little pinprick of light in the upper right-hand corner relative to the size of the black rectangle represents 30 as a proportion of 5 million. The small rectangle in the lower left-hand corner represents the 7,000 substances that have been tested.

These white rectangles greatly exaggerate the extent of existing knowledge. Given the ethical objections to direct testing on humans, most tests for carcinogenicity are done on animals. The report observes "there are no doubt occasions in which observations in animals may be of highly uncertain

Figure 3.5 How much light amid the encircling gloom?

relevance to humans"; it also notes that the transfer of the results of these tests to humans requires the use of scaling factors which "can vary by a factor of 35 depending on the method used", and observes that "although some methods for conversion are used more frequently than others, a scientific basis for choosing one over the other is not established". A further difficulty with most such experiments is that they use high doses in order to produce results that are clear and statistically significant for the animal populations tested. But for most toxins the dose levels at issue in environmental controversies are much lower. Extrapolating from the high dose levels at which effects are unambiguous to the much lower exposures experienced by the general human population in order to calculate estimates of risk for people in the real world requires a mathematical model. The report notes that "the true shape of the dose–response curve at doses several orders of magnitude below the observation range cannot be determined experimentally". It continues "a difficulty with low-dose extrapolation is that a number of the extrapolation methods fit the [high dose] data from animal experiments reasonably well, and it is impossible to distinguish their validity on the basis of goodness of fit". Figure 3.6 illustrates the enormous variety of conclusions that might be drawn from the same experimental data depending on the assumptions used in extrapolating to lower doses. It shows that the estimates produced by the five different models converge in the upper right-hand corner of the graph. Here the five models agree that high dose levels produce high response levels. The supra-linear model *assumes* that the level of response will remain high as dose levels are reduced. The threshold model *assumes* that when dose levels fall below the threshold there will be no response. Below the dose levels used in the experiment one can but *assume*.

There is, in general, no necessary connection between cultural theory type and belief in a particular dose–response relationship. The egalitarian/environmentalist opposition to nuclear power assumes a linear or supra-linear relationship between risk and radiation as the curve is extrapolated into the low-dose region of the graph; there is, they insist, no safe level of radiation. The defenders of nuclear power on the other hand are predisposed, in the absence of conclusive evidence, to adhere to the threshold model, and the belief that the effect on the general public of the radiation they produce is negligible. In the debate about childhood asthma discussed above, the positions are reversed. There has been a large increase in the number of asthma cases diagnosed that are associated with relatively small increases in the suspect pollutants. In order to convict the pollutants, and the car, of being the cause of the increase in asthma, one must invoke a sublinear or threshold type model in which small increases in dose above a certain level produce large increases in response. The defenders of the car will in this case find that the linear or supra-linear models conform better to their expectations.

The argument is further clouded by doubts about the data. Although the

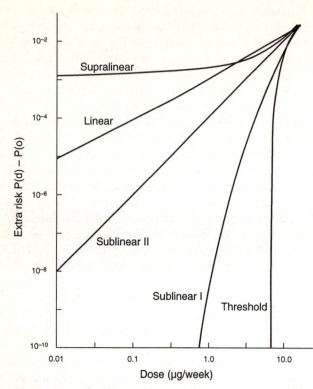

Figure 3.6 Alternative dose–response extrapolations from the same empirical evidence (*source:* National Research Council 1992: 26).

reported incidence of childhood asthma has increased, the asthma death rate for children has actually *decreased* between 1980 and 1990; it is possible that the increase in reported asthma is merely a recording phenomenon, or that the decrease in fatalities is the result of better treatment, or some combination of these explanations. Lenney et al. (1994) conclude that "it is unclear whether the incidence is rising" (see Ch. 5 for a discussion of a similar problem in the interpretation of road accident data.) There are also doubts about the official estimates of trends in pollution, with environmentalists complaining about the sparse sampling on which the estimates are based.

Even amongst those who believe that the increase in childhood asthma is real, there is dispute about the cause. A letter to *The Times* (18 February 1994) from the director of Britain's National Asthma Campaign captures the distinction between individualistic and egalitarian responses to the evidence.

There is a very real danger that, if a general belief develops that asthma is all down to pollution, people will ignore important educational measures about preventive steps that they themselves can take. The scenario of the parents of an asthmatic child who continue to smoke, refuse to give up the family cat, and neglect basic dust-control measures while blaming the asthma on other people's cars, is already worryingly common.

Some go further and blame the increase on excessive environmentalist zeal. Hawkes (1994) notes that energy-saving measures such as double glazing and draught proofing have reduced the number of air changes in houses from as much as four per hour to as little as a half per day – creating an atmosphere in which dust mites flourish.

Beyond the problems of identifying the causes of morbidity and mortality and specifying the dose–response relationships, there are four other sources of uncertainty of even greater significance. First, variability in susceptibility within exposed human populations, combined with the variability in their levels of exposure, make predictions of the health effects of the release of new substances at low dose levels a matter of guesswork. Secondly, the long latency period of most carcinogens and many other toxins – cigarettes and radiation are two well known examples – make their identification and control prior to the exposure of the public impossible in most cases. Thirdly, the synergistic effects of substances acting in combination can make innocent substances dangerous; and the magnitude of the number of combinations that can be created from 5 million substances defies all known computers. And fourthly, the gremlins exposed by chaos theory will always confound the seekers of certainty in complex systems sensitive to initial conditions. After summarizing the difficulties confronting scientists trying to assist the federal regulators of carcinogens, the National Research Council report says "we know still less about chronic health effects other than cancer".

Reports of the National Research Council in the USA, like those of the Royal Society in Britain, carry great scientific authority. The NRC's 1983 report on risk assessment, in keeping with almost all such reports, concludes with a call for more science. The report says:

> The primary problem with risk assessment is that the information on which decisions must be based is usually inadequate. Because the decisions cannot wait, the gaps in information must be bridged by inference and belief, and these cannot be evaluated in the same way as facts. Improving the quality and comprehensiveness of knowledge is by far the most effective way to improve risk assessment.

But when all the unresolved uncertainties discussed in their report are taken into account, the little beacons of scientific light shining in the dark of Figure 3.5 become invisible to the naked eye. The National Research Council's account of what they do not know amounts to an admission that they are groping in the dark. The prospect of future research breakthrough lighting more than a few flickering candles in the vast darkness enveloping the problems they are addressing is not encouraging. Indeed, the problem appears to be getting worse as the rate continues to increase at which chemists, physicists and genetic engineers create new dangers. Even more urgent than the need for more science is the need for a better understanding of the bridge of inference and belief.

The Sydney Smith dilemma

The mythological figures of cultural theory are caricatures, but they have many real-life approximations in debates about risk. Long-running controversies about large-scale risks are long running because they are scientifically unresolved, and unresolvable within the timescale imposed by necessary decisions. This information void is filled by people rushing in from the four corners of cultural theory's typology, asserting their contradictory certitudes. The clamorous debate is characterized not by irrationality, but by *plural rationalities*.

It has probably always been thus. Over 150 years ago the Reverend Sydney Smith was being taken on a conducted tour of an Edinburgh slum. Down a narrow alley between two high-rise tenements he came upon two women shrieking abuse at each other across the alley. Smith stopped, looked up, and listened. He then shook his head and walked on, lamenting, "they'll never agree; they're arguing from different premises".

The enormous gulf between what scientists know or are likely to discover, and what needs to be known before decisions about risk can be based on conclusive evidence, suggests that we are doomed for the foreseeable future to continue to argue from different premises. But the argument is likely to be more civilized to the extent that we are sensitive to these differences and understand their causes. It is, of course, desirable to have as much solid scientific information as possible to inform decisions about risk. There will be occasions when the production of such information will be able to resolve disputes. But for as far ahead as one can see, the future will remain uncertain. The big issues will not be resolvable by science. How then ought we to proceed? How might we manage risk better?

These are questions to which I will return in the concluding chapter. The answers, thus far, appear to depend on whom you ask. The individualist can be relied upon to assert that we are already over-regulated; things should be left to the market to sort out. The egalitarians will invoke the precautionary principle and press for urgent action. The hierarchists will suggest that things are about right as they are, while conceding that more research and a slight nudge to the tiller might be advisable. And the fatalists will carry on watching television and buying lottery tickets.

Chapter 4

ERROR, CHANCE AND CULTURE

"On the technical side, this accident, while no one wanted it, has a statistical probability that falls within the predicted probability of this type of accident."

Chauncy Starr commenting on the accident at Three Mile Island[1]

The conventional wisdom

The implicit assumption of most safety research and safety regulation is that accidents are unwanted.[2] For example, Wright et al. (1988) argue that:

It is a plausible hypothesis that no road user deliberately sets out to have an accident; to put it another way, if it were clear to a road user that a particular course of action would lead inevitably to an accident, he would adopt some other alternative (assuming that one were available).

There is a semantic difficulty here. If it were clear that a particular course of action would lead *inevitably* to an "accident" then the outcome should not be called an accident, because accidents are events that are unwanted and unintended. Defining accidents as unwanted and unintended appears to leave only two possible ways to account for them. Either they are "acts of God" – unanticipatable events beyond the control of the victim – or they are the result of human error – mistakes, misjudgements, lapses of concentration.

Events that are truly unanticipatable and unpreventable lie in a realm of great theological difficulty. Calling them acts of God raises questions about God's intentions, and whether or not He plays dice. We confine ourselves here to the realm of events that are in principle preventable or avoidable by human behaviour, events that hindsight reveals (or could reveal) as preventable.

---

1. Quoted in *Science for People*, 42, Summer 1979.
2. The view of accidents as "Freudian slips" – as the consequence of behaviour subconsciously intended – is a significant exception to this generalization.

Some preventable accidents are doubtless attributable to human error. But are *all* such accidents the result of human error? Most safety research assumes that they are. Summala (1988), for example, suggests that "no consideration is normally given [by motorists] to risks . . . in most situations, drivers obviously know exactly what they should not do if they want to avoid a certain, or almost certain accident". Again we encounter the oxymoronic concept of the inevitable or certain accident. Such terminology is incapable of distinguishing accidental death from murder or suicide.

Summala, like the Royal Society, distinguishes between subjective and objective assessments of danger and argues that accidents are the result of unperceived danger. Risk compensation – such as driving faster if a bend in the road is straightened – Summala describes as "a behavioural response to environmental change". He insists that it should not be called risk compensation because "drivers do not normally feel *any* risk" (my emphasis).

Enter *Homo aleatorius*

Whether or not God plays dice, *Homo aleatorius* does, out of both choice and necessity. As we observed in Chapter 2, no one wants an accident, but everyone appears to want to be free to take risks, and to be his own judge of these risks; a society's accident rate will thus reflect its members' propensities to take risks. Even so, the possibility must be considered that *Homo aleatorius* has more accidents than he bargains for. Might we perhaps hope, not for the elimination of accidents, but for a decrease in accidents by training programmes and improved information to make the gamblers in the casino of life more skilful and better informed?[1] I think not.

Those who consider human error to be the sole or principal cause of accidents advocate safety measures that reduce the likelihood of nasty surprises by signposting dangers, by improving coping skills, or by creating "failsafe" or "foolproof" environments. But this approach is one-sided. It ignores the positive reasons that people have for taking risks –the *rewards* of risk to be found in the upper right-hand corner of the risk thermostat diagram in Figure 2.2.

It is not disputed that some accidents are the result of inaccurate risk assessment. If people underestimate risk, they will have more accidents than they bargained for. However, if it is accepted that people do take risks, then inaccurate risk assessment can cause not only too many accidents *but also too few*. If, as we have argued in Chapter 2, risk-taking is a balancing act, then it is clearly possible to fall off the tightrope in two directions; a risk-taker might take too much risk *or too little*. Figure 4.1 helps to explain.

---

1. Such a suggestion can be found in van der Colk (1988). It is also the principal justification of almost all forms of safety training.

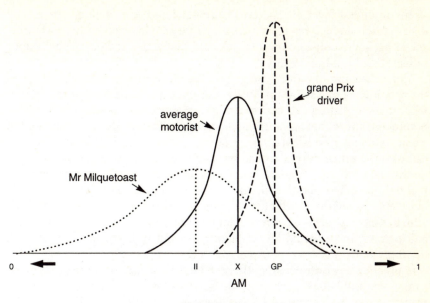

Figure 4.1 "Error" in risk-taking.

The solid line in Figure 4.1 represents a hypothetical risk-taking frequency distribution. In the middle, X, is the intended level of risk-taking, the level represented by the *propensity to take risks* in the upper left-hand corner of Figure 2.2. For an individual who successfully balances perceived and intended risk, X will be the probability of a particular action resulting in an unwanted consequence commonly called an *accident* (lower right-hand corner). If many individuals were to perform the particular action, and the probability of the outcome were to be known, then, although no single action would lead *inevitably* to an accident, these collective actions would lead *inevitably* to a predictable number of accidents. The variance about X will reflect both the number of errors made and the variation in result attributable to chance.

Consider a knowledgeable gambler flipping coins. His intended level of risk is 0.5. If a head is a "success" and a tail is an "accident" and he tosses a fair coin many times, then his predicted number of accidents will be half the number of tosses. His accident rate will equal his intended level of risk. At the end of an evening tossing coins, he would expect his rewards to approximately equal his accident losses. But if the coin were biased, the result would be more or fewer accidents than he bargained for. In practice, for the reasons discussed in Chapter 2, in the uncertainty that exists outside controlled conditions such as those found in casinos, it is not possible to know X, and rarely outside the casino would an individual be able to attach a number to his intended level of risk. So Figure 4.1 is a set of notional frequency distributions embodying the supposition that people live their lives wanting/accepting some non-zero level of risk, and that actual risk-taking behaviour reflects this level imperfectly.

Along the solid line in Figure 4.1, tailing off from X in both directions, is the actual level of risk taken. In discussing distributions such as this, it will be helpful to distinguish between departures from X that are stochastic in nature and those that are attributable to error. The result of a single toss of a coin is usually considered to be down to *chance*. But if you attempt to estimate the probability of a head by tossing many coins, and come up with an estimate different from 0.5, your estimate is in *error* – if the coin is fair. It should be noted that this distinction between chance deviation and error is frequently unclear, with chance often being another name for ignorance, which is closely related to error. If, for example, you were sufficiently skilful and precise, there would be no chance element in the result of your coin tossing.

In practice however, whether through ignorance, error or pure chance, the actual result of behaviour commonly deviates from the intended result in a way that can be described by the frequency distributions illustrated by Figure 4.1. The number of accidents that occur in any particular group is a function of the number of risk-takers and their intended level of risk, plus or minus the "mistakes" they make in matching their behaviour to their intentions. The "balancing behaviour" depicted in Figure 2.2 will rarely be precise; errors in perception and/or behaviour will result in a person taking more or less risk than intended.

To the right of X more risks are taken than desired, and to the left fewer. The variance of the distribution will vary directly with ignorance and ineptitude. The better informed people are about risks, and the more skilful at judging and responding to risks, the smaller will be the variance – the closer they will come more of the time to taking the risks that they intend. If an individual's variance is reduced, there will be fewer occasions when he is exposed to more extreme risk than intended, thereby reducing the chances of an accident; but offsetting this will be the fewer occasions when he is exposed to extreme safety, thereby increasing the chances of an accident. A Grand Prix racing driver, for example, will be better informed about the dangers associated with the track and his car, a better judge of driving risks, and a more skilful driver than the average motorist – but not less likely to have an accident. He will use his superior skill and judgement to convert some of the average motorist's safety margin into increased performance. If, as is likely, he has a higher than average propensity to take risks – for which there are lucrative rewards – he will be likely to have more accidents. Williams & O'Neil (1974) found that drivers with specialist training for off-highway racing had worse on-highway accident records than ordinary drivers; they were more skilful drivers, but not safer. If we let the solid line represent the distribution of errors by the average motorist (AM), the dashed line, with its higher level of intended risk and smaller variance (GP), might represent the distribution of errors by a Grand Prix racing driver.

This distinction between level of risk and error is at odds with Mary Douglas's (1986) view of risk.

> The essence of risk-taking lies in the structure of the probabilities, their variance. A prudent individual seeks less, the risk-taker prefers more variance. A theory of decision-making that takes the mean of the distribution of probabilities disregards the very thing that risk-taking is all about, the distribution itself.

This view cannot distinguish between skilful takers of high risks, such as the Grand Prix driver – judging with great precision just how fast he can take a corner – and ignorant, inept prudence. *On average* the ignorant and ineptly prudent (II) will have fewer accidents (dotted line in Fig. 4.1). They will on occasion stray into great danger, but these occasions will be more than offset by those occasions in which their caution leads them to be excessively careful.

Excessive prudence is a problem rarely contemplated in the risk and safety literature. Cases of accidents resulting from ignorance or incompetence are numerous, and well documented in accident reports. There are literally hundreds of journals devoted to the examination of accident statistics with the aim of reducing accidents. But there are also many much less well documented examples of people taking *less* risk than they desire through ignorance or incompetence. It is widely ignored because, from the perspective of those seeking to increase safety, it is not a problem. A few examples will serve to indicate the pervasiveness of this neglected phenomenon:

- Overestimates of risk can lead people to spend more on insurance than they otherwise would.
- Motorists drive very slowly if they believe, falsely, that there are patches of black ice on the road.
- In the construction industry excessive prudence can lead to an enormous waste of money if buildings are designed for stresses with which they are unlikely to have to cope: applying earthquake zone standards in areas unlikely to experience earthquakes, for example.
- On the railways in Britain, a spate of recent accidents has arguably led to excessive safety, excessive in the sense that the new safety measures will be paid for by fare increases that will encourage some passengers to travel by more dangerous cars instead.
- Inordinate fear of mugging or physical attack leads some women and elderly people to confine themselves to quarters and deny themselves freedoms that they would otherwise enjoy.
- A personal example. I still have a vivid childhood memory of excessive safety leading to social isolation. I refused, through days of agonizing, to

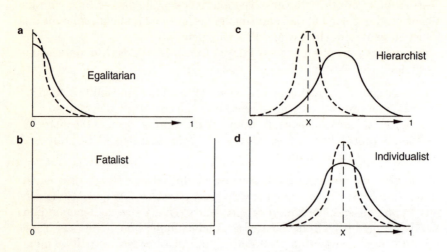

Figure 4.2   Error cultures.

play with friends in a favourite haunt until it was explained to me that the rusty sign "Trespassers will be prosecuted" did not mean "electro-cuted".

Again it must be stressed that because of intractable measurement prob-lems it is not possible to attach numbers to the speculations embodied in Fig-ure 4.1. They are analogies, applications of statistical concepts to a problem that cannot be reduced to numbers. *All* attempts to formalize and quantify the making of decisions about risk are fragile vessels afloat on the sea of uncertainty; even in the casino one might doubt the honesty of the staff and management. As Chauncy Starr demonstrates in the quotation at the begin-ning of this chapter, it is not possible to formulate falsifiable statements about unique future events in terms of probabilities. If God does play dice, then the attempts of mere humans to attach probabilities to the outcome of the celestial craps game will always be laughable. Nevertheless, the *intended level of risk* and *the variance about this level produced by chance and error* are concepts that it is important to separate in any attempt to understand risk-taking behaviour.

Balancing behaviour

Having separated these concepts and related them to the idea of risk as a balancing act called risk compensation, let us now speculate further with the help of Figure 4.2. It illustrates the way in which cultural theory helps to account for different settings of the risk thermostat and different styles of balancing act.

Figure 4.2a illustrates the intended level of risk and the variance implicitly assumed in most of the safety literature. For example in *Human error*, a recent comprehensive work on the topic, Reason (1990) does not consider either chance or deliberate risk-taking as significant causes of accidents. He advances the "basic premise . . . that systems accidents have their primary origins in fallible decisions". The level of risk intended by most of those in charge of safety, and by the researchers whose work they fund, is zero in Figure 4.2a. There is only one direction in which one can fall if one loses one's balance – in the direction of greater than desired risk. This perspective on risk is characteristic of *egalitarians* who adhere to the myth of nature ephemeral. Because the potential consequences of error are so enormous, they strive unrelentingly to reduce the variance (to move in the direction of the dotted line), and hence the risk of things going wrong; it is sometimes acknowledged that zero risk is an unattainable ideal but, nevertheless, one towards which we should all continually strive. Those who believe it is actually attainable are clearly deluded.

The flat frequency distribution of Figure 4.2b, ranging from zero to one, characterizes the perspective of *fatalists*. Nature is simply unpredictable. One variant of fatalism holds that all is predestined, another that God throws dice. But ignorance precludes adherents to either perspective knowing what the future holds. As fatalists they are entitled to no intentions with respect to risks, only hopes. They can but hope for the best – and duck if they see something about to hit them.

Figure 4.2c represents the *hierarchist* style of risk-taking. The solid line represents their conviction that those under their authority, persistently have more accidents than they should. They seek to reduce risk. They usually concede the impossibility of reducing it to zero, but seek to manage it more efficiently. Implicit in their attempts to manage risk better are two beliefs: first, that through ignorance or incompetence people persistently take higher risks than they intend, with the result that the number of accidents is greater than that implied by the accepted risk level X; secondly, that many people under their authority are irresponsible and accept higher levels of risk than they should. The hierarchist adopts a paternalistic approach to risk regulation; not only must people be dissuaded or prevented from behaving in a way that puts other people at risk (as in campaigns against drunken driving), they must also be protected from themselves (as in seat belt legislation). Sometimes they resort to exclusion. One line of safety research going back many years seeks to identify the accident prone. A recent variation on this theme is research in pursuit of "hazardous thought patterns" (Holt et al. 1991). The Swedish airforce tries to identify such patterns with its Defence Mechanism Test, and rejects any aspiring pilot who fails the test (Dixon 1987). Hierarchists strive through engineering measures, persuasion, regulation, training and exclusion to shift the frequency distribution of risk-taking behaviour to the left and reduce its variance (dotted line).

Figure 4.2d represents individualists. They also seek to reduce their variance (dotted line), and are assiduous collectors of information about risk – whether it be on the race track or the stock market. But they are more alert to the *rewards* of risk-taking. They are self-conscious risk-takers and they trade slogans such as "no pain, no gain" and "no risk, no reward" and are convinced that a benign nature will ultimately reward those who trust her. They trust individuals to make their own decisions about risk and scorn the regulators of the nanny state.

Types of error

These different styles of risk-taking can be related to the type 1 and type 2 errors of the statistician. A type 1 error is committed if one accepts the hypothesis that one ought to reject, and a type 2 error is committed when one rejects the hypothesis that one ought to accept.[1] The statistician's "confidence level" is a measure of the risk of error. The 95 per cent confidence level most commonly employed in social science research means that the researcher accepts the probability of getting it wrong one time in twenty. The four myths of nature are contextual hypotheses constantly being subjected to partial tests. Consider the specific hypothesis that CO_2 emissions threaten a runaway greenhouse effect. Egalitarians whose working hypothesis (myth of nature) states that catastrophic consequences will flow from a failure to respect the fragility of nature will insist on a very high standard of proof before rejecting this hypothesis; in the statistician's language he will be prepared to run a high risk of a type 1 error and a low risk of type 2. Conversely individualists who are convinced of the robustness of nature will require a very high standard of proof before accepting the hypothesis. The hierarchist who believes in stability within limits will return the hypothesis to the sender, requesting greater specificity with respect to critical limits. Disputes amongst adherents to these different perspectives usually turn out to be arguments not about "facts" but about where the burden of proof should lie.

The field of risk and safety research is dominated by the concern to *reduce* risk and accidents. The two risk-taking stereotypes that share this concern are egalitarian and hierarchist. The egalitarians are usually more risk-averse than the hierarchists. The hierarchists are usually responsible for putting safety measures into effect. They commonly find themselves lobbied from two sides, with the egalitarians urging more action to reduce risk, and the individualists insisting on less. The fatalists see no point in arguing.

---

1. Strictly, a type 1 error is committed by wrongly rejecting the null hypothesis, and as a consequence provisionally accepting the hypothesis that is the converse of the null hypothesis, and a type 2 error is committed by wrongly accepting the null hypothesis.

Acceptability of risk

There are long-running arguments in the risk literature about what risks and levels of risk are *acceptable* (see, for example, Douglas 1986 and Fischoff et al. 1981). Hierarchists, egalitarians and individualists are all participants in these arguments, and the arguments will continue to run because the participants are arguing from different premises. At one extreme are those who argue that one death is one too many, and at the other those who interpret the prevailing accident rate in areas of voluntary risk-taking (about which more shortly) as a measure of the level of risk that society as a whole finds acceptable. In between are those who argue, not very specifically, for less risk. The *behaviour* of the hierarchists and egalitarians in debates about safety policy can be considered a form of risk compensation; their striving to reduce risk for the general population implies that the danger they perceive is greater than the risk they consider acceptable. And yet ironically most of the risk-reducing measures they propose and implement deny the existence of this phenomenon in the people on whose behalf they would claim to legislate.

The efficacy of intervention

Because people compensate for externally imposed safety measures, the risk regulators and safety engineers are chronically disappointed in the impact that they make on the accident toll. In most countries in the developed world the rate at which new safety regulations are added to the statute book greatly exceeds the rate at which old ones are removed. Although a few libertarians have railed against the excesses of the nanny State, the preponderance of political pressure over many decades has been on the side of more State interference to reduce risk. There is an even greater imbalance in the area of research. Huge sums of money are spent on safety research; although there is considerable research in the economic realm associated with individualist and market-based ideologies, minute sums are spent on countervailing research in the realm of physical risk and safety. What has been the effect of this long-term imbalance?

Figure 4.3 show indices of death by accident and violence for 31 countries over the first 75 years of the twentieth century. The indices are standardized mortality ratios; this means that the effect of differences between countries, or over time, resulting from differences in age and sex distributions have been removed. The indices show averages for periods of five years, so the last value shown for each country represents the average standard mortality ratio for the period 1971–5.

Interpreting data covering such a long time and so many different countries is notoriously difficult. The data cover a period in which the international conventions for classifying causes of death underwent several

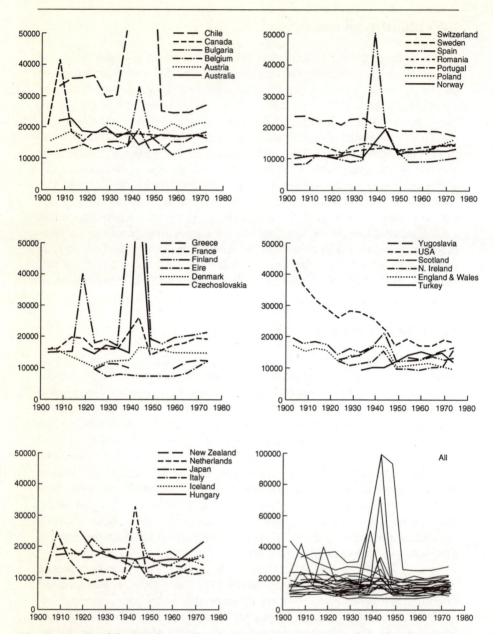

Figure 4.3 Death by accident and violence. Standardized mortality ratios for accidental and violent death for 31 countries between 1900 and 1975. *Source:* Adams 1985.

changes. The quality of the data can be assumed to vary over time and between countries; for example, the indices include suicide, a cause of death believed to be variably under-recorded according to the stigma attaching to it in different countries. However, with few exceptions, the data exhibit no clear long-term trend – there is a slight upward drift after the Second World War, during a period when safety efforts in most countries intensified. But I am aware of no source of recording bias common to all 31 countries that could mask a falling trend.[1] The 75-year period covered by Figure 4.3 is a time over which great improvements were made in casualty treatment. Over this period, all countries conducted many inquests and safety inquiries, passed volumes of safety regulations, and appointed small armies of safety regulation enforcers. But the graphs of rates of accidental and violent death remained remarkably flat – with the exception of marked spikes associated with wars or very large natural disasters.

It should be noted that this flatness does not appear remarkable to some historians and demographers. In a survey of death by accident and violence in Britain since the thirteenth century, Hair (1971) reports changes over time in the particular causes of accidental and violent death, but no apparent trend in the rates for all causes aggregated together. Although rates in 1970 were below mid-nineteenth-century rates, they were higher than estimates for most preceding centuries. He concluded

> British society throughout the centuries has struggled to control vio-
> lence, and has frequently succeeded in taming one form – only to find
> another emerging. The axe of the drinking companion and the neigh-
> bour's open well were regulated, to be replaced by unruly horses and
> unbridged streams; when these were brought under control it was the
> turn of unfenced industrial machinery and unsignalled locomotives:
> today we battle with the drinking driver.

And in a demographic study projecting English and Welsh mortality rates into the 21st century, Benjamin & Overton (1981) construct several scenarios. Their "optimistic" scenario incorporates the assumption that up to the year 2018 "the risk of accidental death remains the same, as some of the improvements in the environment are balanced by the appearance of new hazards".

Can Figure 4.3 be interpreted as support for the view that risk compensation has been taking place on a societal scale with invariant risk-taking propensities over a very long period of time? Does it constitute support for the individualist position that the accident outcome is a measure of risk acceptability? Perhaps. Judging by these statistics, risk appears to have been suppressed in some activities only to pop up in others. Certainly there appears to be little to show in the aggregated statistics of death by accident and violence for all the labours of the risk reducers – the regulators, the

---

1. The statistics for different countries suggest that fatalities attributable to war are not treated consistently. For Japan the series is simply broken during the Second World War. The statistics for England and Wales appear not to contain war fatalities, although those for Spain, Finland, Czechoslovakia and the Netherlands appear to include them.

police, the doctors, the safety engineers and all the others involved in the safety industry over many decades. The casualty rate associated with the dance of the risk thermostats appears to have been remarkably little perturbed by their activities. During the 20th century dance, some of the tunes and the dance steps have changed, old dancers have left the floor and new ones have arrived – fewer people are trampled by horses and more are killed by cars – but the overall level of mayhem, accidental and intentional, continues unabated at levels that display no trend up or down.

But there are certain difficulties with interpreting this lack of trend as evidence for risk compensation. The risk compensation hypothesis is an explanation of individual, not collective, behaviour, and there is nothing in the hypothesis that requires either the propensity to take risk or the perception of danger to be constant over time. Further, the multi-dimensionality of risk and all the problems associated with measuring it discussed earlier, preclude the possibility of devising any conclusive statistical tests of the hypothesis.

Even death passes through cultural filters. During the past 25 years of "The Troubles" in Northern Ireland, for example, every death attributed to terrorism has received great publicity, but only a few dedicated collectors of statistics are likely to be aware that over this period twice as many people were killed in road accidents. A terrorist murder conveys powerful messages; some lives are expendable, some invaluable. From one side of the sectarian divide come cheers for the perpetrators, from the other come vows of revenge. The forces of law & order react to the challenge to their authority. Those uncommitted to either side of the struggle deplore its "irrationality" and resent the interference in their lives of both the terrorists and the heavy hand of the security services. Perhaps only for the true fatalist is the random sectarian killing comparable (in the scales in which loss is measured) to the equally meaningless fatal road accident. For the rest, there are no units in which the rewards and losses of traffic and The Troubles can be measured.

People take or have imposed upon them many risks not related to Figure 4.3. They die of causes other than accident and violence, and the importance of some of these other causes has diminished greatly in relative importance over time. Figure 4.4 shows the remarkable progress that has been made in reducing the effects of infectious diseases; over the same period the lack of progress, at the aggregate level, in dealing with accident and violence has greatly increased the relative significance of the latter as a cause of death for people below the age of 40. For men aged 20, in 1931 infectious diseases accounted for about 36 per cent of fatalities, and accident and violence only 20 per cent. By 1982 infectious diseases accounted for less than 2 per cent, and accident and violence for about 70 per cent.

Medical risks are difficult to compare with the risks of accidents and violence, because they tend to operate more slowly and their diagnoses can be contentious. Most of the causes of death by accident and violence can be

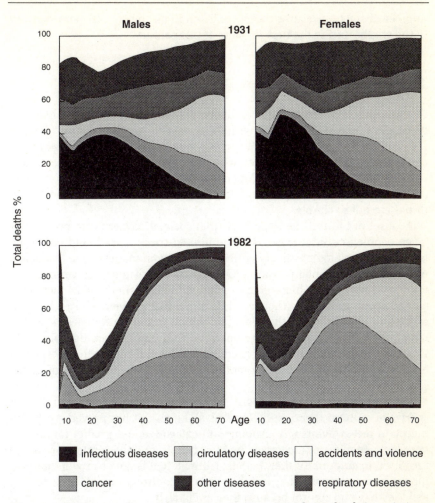

Figure 4.4 Causes of mortality by age (*source:* British Medical Association 1987).

distinguished from most of the other causes of death by the greater speed with which cause leads to effect, and by the greater clarity and intelligibility of the connection between cause and effect. The established international convention used for classifying road accident fatalities attributes a death to a road accident only if the victim dies within 30 days of the accident; after that time it is attributed to a complication – pneumonia, kidney failure, and so on. Where the relationship between cause and effect is unclear and long delayed, as with smoking and lung disease or radiation and cancer, there are few opportunities for risk thermostats to function. Nevertheless, if there has been no collective turning down of the setting of risk thermostats, the pattern of the graphs in Figure 4.3 is what one would expect to find, despite the large number of safety measures that were taken in all the countries represented in the graphs during the first 75 years of this century.

The importance of context

The four-part typology of cultural theory has been previously presented as a typology of "ways of life", each with its associated "myth of nature". People, the Cultural Theorists have argued, cling tenaciously to one of the four ways of life unless and until confronted with overwhelming evidence of its lack of viability, at which point they adopt one of the other three. But work by Dake (1991) on "orienting dispositions in the perception of risk" suggests that things are not so simple.

Dake set out to test the hypothesis that societal concerns are predictable given people's cultural biases. He had limited success. He found that egalitarianism correlated positively with concerns about matters such as nuclear war and environmental pollution. On such issues egalitarians were notably more risk averse than hierarchists or individualists. Hierarchists were concerned about threats to authority, while individualists were worried by over-regulation and threats to the economy, which forms the framework within which their free-market ideology operates. And both hierarchists and individualists were much less concerned than egalitarians about the prospect of nuclear annihilation, but much more concerned about the threat of Soviet expansion.

Although Dake found statistically significant correlations between the categories of cultural theory and concerns about risk, it has proved a difficult theory to test convincingly. Attempts to categorize and predict founder in tautology; people are categorized by their beliefs, and these categories are in turn used to account for their beliefs. Ultimately it is not clear whether people are fatalists because they feel they have no control over their lives, or they feel they have no control because they are fatalists.

And although Dake's analysis yielded several correlations between cultural bias and concern about risk that were statistically significant and consistent with hypotheses generated by cultural theory, the correlations were weak and of limited predictive value; the strongest was 0.46, indicating that cultural bias could account for only just over 20 per cent of the variance in concern about nuclear war found amongst his sample of egalitarians. Most of the other correlations could account for considerably less of the variance. Dake concludes that:

> The perception of risk is set in a continuing cultural conflict in which the organization of rights and obligations, and of community loyalties and economic interests, matter at least as much as scientific evidence.

But it must be conceded that attempts thus far at verifying the speculations of cultural theory with quantifiable data have been very partial successes. Or, put another way, insofar as it has been possible to formulate testable hypotheses from the theory, it appears to offer only a partial explanation of the variance observable in people's responses to risk.

It is common ground shared by psychology and anthropology that the world is experienced through filters that are the product of earlier experience. Disciplinary biases lead psychologists to focus on the unique character of these filters and anthropologists to generalize about their social origins and common features. In Chapters 2 and 3 we noted the uniqueness of every risk thermostat, and concluded that there are as many frames of reference in the world as their are risk-takers. But unless one can find some patterns in all this uniqueness there is little further to be said.

Scale, and voluntary versus involuntary risk

Consider a case where cultural theory throws up a paradox. Greenpeace campaigners will take enormous personal risks in their flimsy rubber boats in order to reduce risks at a planetary level. Are they individualists or egalitarians? Can they be both at the same time? The answer appears to be bound up with issues of scale and voluntariness.

Two different risks are being addressed simultaneously in the Greenpeace example, and in both cases the risk thermostat can be seen to be in operation. At the personal level the perception of the rewards of risk (saving the whale or the world, and perhaps the glory that attaches to such an achievement) lead to a high propensity to take personal risks. In the taking of the actual risk in the rubber boat, vigilance levels are high, and the balancing act requires great skill. The daredevil attitudes and abilities required invite comparison with the Grand Prix driver. They are voluntary and individualistic in character. The myth of nature commonly associated with such activities is nature benign – the risk is taken optimistically in the expectation that the gamble will pay off.

But the action of the Greenpeace boatmen ostensibly addresses another risk, much larger in scale and involuntary in the sense that it is perceived as imposed by others. This larger risk, to the whale or the world, threatens something of value to the group to which they belong, and collective action is taken to reduce it. The activity in the rubber boat is the culmination of a host of group activities involving political action, fund-raising and buying boats. The racing driver on the other hand, although also dependent on the support of a skilled team, is motivated by rewards that are selfish; he performs in a milieu that applauds the pursuit of self interest, and evinces little concern for the wider environmental impacts of its activities.

Several studies have attempted to distinguish between voluntary and involuntary risk, and have concluded that people are prepared to accept

65

much higher levels of risk from activities that are voluntary. Starr, for example, reported that the public is willing to accept risk from voluntary activities such as skiing, which are roughly a thousand times greater than those it will tolerate from involuntary activities that provide the same level of benefit (reported in Fischoff et al. 1981).

Several difficulties arise from such contentions. First there are the problems already discussed about measuring risk; how, for example, does one compare the risks of skiing to those of nuclear power stations? Secondly, there are equally intractable problems encountered in measuring benefits; how does one compare the benefit of skiing with the benefit of electricity from a nuclear power station. Thirdly, how does one measure voluntariness?

Risk compensation and cultural theory cast some helpful light on these questions. The dance of the risk thermostats suggests that there are degrees of volition in the taking of risk, depending on the relative sizes of those imposing the risk and those imposed upon; the greater the relative size of the person or agency imposing the risk, the less voluntary the risk will appear to those imposed upon. But the perception of voluntariness will also depend on where the threat comes from. If it comes from within one's own culture, it will not be seen as imposed. To a supporter of nuclear power, for example, the risks associated with the industry will be willingly borne, while to an opponent they will feel imposed; the former will place a higher value than the latter on the rewards of nuclear power. The perception of risk will also be affected; fewer dangers are likely to get through the supporter's cultural filter. Natural hazards, represented by lightning and the Beijing butterfly, will either be viewed fatalistically, if nothing can be done about them, or considered voluntary, if exposure to them can be controlled. Acceptability of risk also has an economic dimension. Poverty will affect the perception of rewards and dangers and can induce people to take extra risks. There is a steep social–economic class gradient to be found in accident rates, with the poorest experiencing much higher rates than the wealthiest. The concept of "danger money" is sometimes explicit in the pay structures of dangerous industries. The fact that there is no clear downward trend to be seen in Figure 4.3, despite the fact that all the countries represented experienced great increases in affluence over the first 75 years of this century, provides some support for the relative income hypothesis that maintains that, above subsistence levels, it is relative not absolute income that spurs people on.

Despite limited success in validating the cultural theory typology statistically, it remains useful; but it is necessary to modify its application. The four myths of nature are partial truths; each is valid some of the time in some places. People clearly vary their risk-taking according to circumstances. The variability found in the risk-taking behaviour of an individual cannot long be contained within a single one of the four "ways of life" depicted in Figure 3.3. The same person might, depending on context, behave as an egalitarian supporting environmentalists in their campaign to save the whale, as a hierarchist campaigning for more government regulation of the motor car or the stock market, as an individualist resentful of the law that requires him to wear a motor cycle helmet, and a fatalist on receipt of the news that he has cancer.

"Ways of life" implies a stability, consistency and comprehensiveness of value systems that is difficult to find in a pluralistic world. Indeed, it has proved impossible to find pure examples of the cultural theory stereotypes. Breathes there a person outside the asylum who has *never* made common cause with others to achieve some objective, or who acknowledges *no* social constraints on his behaviour, or in whom *all* traces of individuality have been extinguished by subservience to authority, or who is *not* rendered fatalistic by the contemplation of his own mortality? A more flexible term is needed; cultural theory is better understood not as a set of categories of "ways of life" but as a *typology of bias*. Some cultures are more individualistic than others, some more hierarchical, some more egalitarian and some more fatalistic. But in certain circumstances risk can draw out each of these tendencies in *any* culture.

Error, chance and culture

Insurance companies sell policies at a price that they hope will cover future claims and leave them a reasonable profit. Their anticipations can be represented by a frequency distribution such as the solid line Figure 4.5. The claim rate for the average policy is ACR. To the right are the bad risks, to the left the good risks. Sometimes they disaggregate their policies, charging higher rates to insure young male motorists, or lower rates for non-smokers, for example. They know that within the population that purchases their policies there are other groups that will have higher or lower than average claim rates, but the costs of identifying them accurately, devising sales campaigns to reach a segmented market, and designing separate charging schedules for them, are considered not worth the effort. They rely on the law of large numbers to balance the bad risks with good ones.

If a policy holder submits a claim it could be a consequence of a high risk deliberately taken, or of ignorance or incompetence, or it could be from someone who was careful but unlucky. The solid line in Figure 4.5 conceals variation attributable to error, chance and culture (dotted lines). Any approach to risk that does not acknowledge the rôle of *error* and *chance* and *culture* in shaping attitudes, influencing behaviour and determining outcomes will be inadequate for coping both in the insurance industry and in the casino of life.

Consider again the dance of the risk thermostats. However big and powerful you are, there is almost always someone bigger. However small and insignificant, there is almost always someone smaller. There are different, competing bands in each corner of the floor, playing different tunes with different

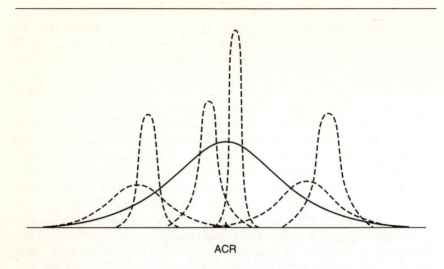

ACR

Number of Claims

Figure 4.5   Good risks, bad risks.

rhythms. The dancers form clusters; some prefer formation dancing, others individualistic jiving, some have flailing arms and legs and are given a wide berth by others, some are wall flowers lurking on the margins, some will loosen up after a drink or two. Some move about the floor, others tend to stay put. All human life is there, but no one on the dance floor can possibly have more than a partial view of what is going on. Risk compensation and cultural theory provide a precarious imaginary vantage point above the dance floor, discern motives and pattern in all this activity. They provide a conceptual framework for making sense of this ever-changing order in diversity, and a terminology with which people can discuss how best to cope with it.

Chapter 5

MEASURING RISK

La révolution est un bouleversement qualitatif des statistiques.
From "The Sayings of President Sankara of Burkina Faso"[1]

Risk, according to the definition most commonly found in the safety literature, is the probability of an adverse future event multiplied by its magnitude. We have already noted the elusiveness of objective measures of risk;
records of past accidents are not trustworthy guides to the future because
people respond to risks as soon as they see them, and thereby change them.
But, given the still deeply entrenched view that accident statistics are a useful measure of safety and danger, and given that they are still virtually the
only measure used for assessing the success of safety interventions, I turn
now to two further problems with these measures. The first is the unreliabil-
ity of the historic accident record; not only is it an untrustworthy guide to
the future, it is an untrustworthy guide to the past. The second is the absence
of an agreed scale for measuring the magnitude of adverse events.

Cultural theory can shed light on both these problems. It suggests that risks
are viewed through cultural filters; the institutional arrangements for monitoring risks through the collection and analysis of statistics relating to mortality, morbidity, economic damage, and near misses will all reflect the
biases of the collectors and analysts.

Not enough accidental deaths

For the risk analyst, death has two great attractions. It is the least ambiguous
of all the available measures of accident loss and, because it is usually
considered the ultimate loss, it is the most accurately recorded. Deaths are,
however, sufficiently infrequent, and their causes sufficiently diverse, to
make them an unreliable guide to remedial action. Any analysis of the causes
of accidents always leads to the conclusion that they are a *stochastic* or

---

1. I am grateful to Andrew Warren for spotting this on a billboard in Ouagadougou.

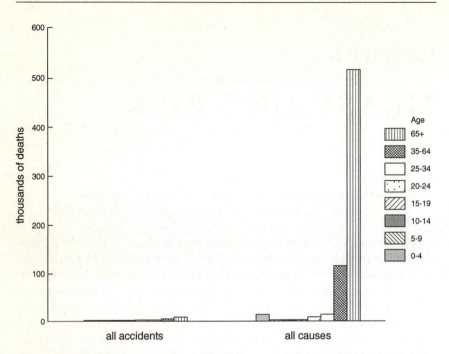

Figure 5.1   Accidents as a cause of death (*source: Road accidents Great Britain 1992*, Department of Transport).

probabilistic phenomenon. They result from a conjunction of circumstances to which the victim had, *a priori*, in most cases assigned a negligible probability. In the case of *fatal* accidents, the probabilities are very low indeed: in 1991 in Britain, in a population of 57 million, there were 12,816 accidental fatalities out of a total of 628,000 deaths by all causes. Figure 5.1 illustrates how, in terms of absolute numbers, fatal accidents barely register on a graph alongside all causes of death.

These 12,816 accidental deaths were spread over hundreds of thousands of kilometres of roads, millions of motor vehicles and billions of vehicle kilometres, plus countless boats, ladders, stairs, rivers, windows, swimming pools, guns, knives, electrical appliances, inflammable materials, medicines and toxic substances, to name but a few of the more obvious hazards. Every year in Britain alone there are many billions of *potentially* fatal events that imagination might construe out of all these hazards. The problem for the safety planner is that the connection between potentially fatal events and actual fatal events is rendered extremely tenuous by the phenomenon of risk compensation. Precisely because events are recognized as potentially fatal, they are rarely so; avoiding action is taken, with the result that there are insufficient fatal accidents to produce a pattern that can serve as a reliable guide to the effect of specific safety interventions.

As a consequence, safety planners seek out other measures of risk: usually – in ascending order of numbers but in descending order of severity – they are *injury, morbidity, property damage* and *near misses*. Where numbers of accidents are small, the accident statistics commonly display great variability, both temporally and geographically. The ratio of all reported injuries to fatalities is usually large – for road accidents in Britain about 70 to 1, and it is much easier to achieve "statistical significance" in studies of accident causation if one uses large numbers rather than small. Safety researchers therefore have an understandable preference for non-fatal accident or incident data over fatality data, especially when dealing with problems or areas having few accidents. In exercising this preference, they usually assume that fatalities and non-fatal incidents will be consistent proportions of total casualties, and that the larger, non-fatal, numbers provide the best available indicators of life-threatening circumstances.

Figure 5.2 casts doubt on this assumption. It shows that, measured by road accident injury statistics, in 1985 London was the most dangerous jurisdiction in Britain with 759 injuries per 100,000 population. But, in company with most of the other English conurbations, London had one of the lowest recorded death rates (7.3 per 100,000). The average injury:fatality ratio for Britain cited above of 70:1 conceals a wide range – from 103:1 for London down to 23:1 for Dumfries and Galloway.[1] The correlation between fatality rates and injury rates is very weak. Figure 5.2 raises two interesting questions for those who use casualty statistics as a measure of risk. First, is the weak correlation between injuries and fatalities real or simply a recording phenomenon? Secondly, how many injuries equal one life?

There are two contending explanations for the pattern displayed in Figure 5.2. The first is that it reflects real differences in driving conditions. For every mile travelled on a road in a built-up area (a road with a speed limit of 40 mph or less) in Britain in 1991 a motorist had a 127 per cent greater chance of being *injured* than he would if travelling on a road in a non-built-up area (with a higher speed limit). Does this mean that the roads with lower speed limits are more dangerous for motorists? Not necessarily. The same source of statistics (Department of Transport 1992) suggests that the chance of being *killed* per mile travelled is 12 per cent higher on the roads with the higher speed limits. The ratio of injuries to fatalities on roads in built-up areas is 98:1 and in non-built-up areas is 39:1. Thus, at least a part of the explanation for Figure 5.2 appears to lie in the fact that London is so congested and traffic speeds so low that there are large numbers of minor accidents, but that high-speed crashes resulting in more serious damage are relatively rare. Conversely, on the rural roads of Dumfries and Galloway there is much less traffic, but it travels at speeds that make the consequences of accidents, when

---

1. The ratio for the Isle Wight in 1985 was 140:1, but this is based on a fatality figure which is very small (5) and unstable from one year to the next.

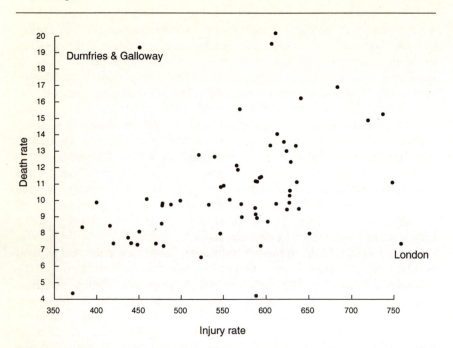

Figure 5.2 Road accident death and injury rates in Great Britain per 100,000 population. *Source:* Adams 1988.

they do occur, much more serious.

So which class of road, or area, is safer? How many injuries equal one life? Is London the most dangerous part of Britain, or one of the safest? If safety measures could be implemented, which slowed traffic on the road with the result that the injury rate increased but the fatality rate decreased by a lesser amount, could the measures be described as a safety success? Whose safety record is better, Dumfries and Galloway's, or London's? Safety measures such as straightening bends in roads, lengthening sight lines, improving cambers or raising the coefficient of friction of road surfaces could all have the effect of reducing numbers of accidents, but also of increasing speeds and the number of *fatal* accidents.

A further uncertainty arises from the fact that most road safety measures have highly localized effects. In a small area, or at the site of a treated accident black-spot, a decrease in the number of injury accidents might be "statistically significant", while a "real" increase in fatalities could remain statistically undetectable. And even if the effects of a safety measure on both injuries and fatalities could be detected with confidence, until these two measures of risk can be measured on a common magnitude scale there can be no objective way of deciding which accident record to prefer. But before contemplating this preference further, we must first consider a second possible explanation for the difference between the accident records of London and Dumfries and Galloway: that it is not a real difference at all, but an arte-fact of the way the statistics are collected.

What gets recorded?

The government are very keen on amassing statistics. They collect these, raise them to the nth power, take the cube root and prepare wonderful diagrams. But you must never forget that every one of these figures comes in the first instance from the village watchman who puts down what he damn pleases.   Sir Josiah Stamp (quoted in Nettler 1974)

The problem of the reliability of data that we are discussing is not new. Figure 5.3 and Sir Josiah together suggest a second explanation for the variation displayed in Figure 5.2, or at least for part of it. Injuries are under-recorded – variably, substantially and, almost certainly, systematically. Between 1930 and 1993 the number of people *killed* in road accidents in Britain has *decreased* by 48 per cent (from 7,305 to 3,814). Over the same period the number of recorded road accident *injuries* has *increased* by 72 per cent (from 178,000 to 306,000). Since 1930 there have been improvements in rescue services and the medical treatment of crash victims, and cars have become more crashworthy; so perhaps the increase over time in the ratio of injuries to fatalities is real. But cars have also become much more powerful and faster, and lorries have become much heavier, with the result that the phys-ical damage they can cause in a crash is much greater. Further, by 1992 there were more than twice as many police in Britain as in 1930. So perhaps the change is simply the result of a larger fraction of injury accidents being recorded now than in 1930. It is also possible that at least part of the geo-graphical differences in the injury:fatality ratios might be accounted for by a higher degree of under-reporting of minor injuries in more sparsely popu-lated areas where police are thinner on the ground. London has almost twice as many police per 1,000 population as the rest of Britain. One does not need to go very far in London to find a policeman in order to report a minor acci-dent.

The "Severity Iceberg", Figure 5.3, shows the way in which uncertainty in the data increases as the severity of injury decreases. The fatality statistics are almost certainly the most accurate and reliable of the road accident statistics. Death on the road in most countries is treated very seriously. Sev-eral studies from countries all around the world (summarized in Hutchinson 1987) which have compared police and hospital statistics have found that virtually all fatalities recorded by hospitals are also recorded by the police. For injuries, however, the situation is less satisfactory. In most countries the classification of injuries is done within a short time after the accident by medically unqualified police. The categorization and recording of injuries is

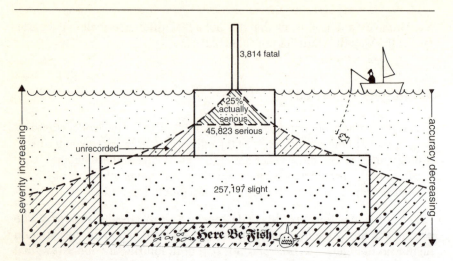

Figure 5.3 The Severity Iceberg: the areas of the rectangles are proportional to the numbers of casualties recorded in *Road accidents Great Britain 1993 (source:* Department of Transport 1994).

generally not informed by any evidence from a medical examination. In Britain, according to evidence given by the British Medical Association (1983) to the House of Commons Transport Select Committee, the resulting numbers are not only defective, but positively misleading. They said

> The existing definitions on which records are based are misleading. Only one in four casualties classified as seriously injured are, in fact, seriously injured and many of those classified as slightly injured are in fact seriously injured. The existing [police] definition of "seriously injured" covers everything from a broken finger to total paralysis and to death occurring more than 30 days after the accident. Within these unsatisfactory definitions there is wide-spread under-reporting and mis-reporting of casualties and the distribution of these errors varies widely between different categories of road user. The information is very defective in the case of pedestrians and cyclists, who are at high risk of serious injury as a result of their lack of protection.

Curious about how medically untrained police might cope with the distinction between "slight shock" and "severe general shock" – one of the criteria by which slight accidents are distinguished from serious accidents – I asked some policemen that I encountered in the street in London. None of them was very sure. One of them, trying to be helpful, recalled a case where an electricity cable had fallen across a car and electrocuted the occupant; that, he was sure, was serious shock.

The BMA went on to present evidence that some 30 per cent of traffic accident casualties seen in hospital are not reported to the police, and that at least 70 per cent of cyclist casualties go unrecorded. It is not known how much of the variation in the injury:fatality ratios displayed in Figure 5.2 is real and how much is the result of variation in recording practice. The BMA also noted that the degree of under-reporting increases as severity decreases. Towards the bottom of the Severity Iceberg under-reporting will approach 100 per cent; there will be a degree of severity which is sufficiently slight that neither the injured person nor the police will consider it worth reporting.

It is widely accepted that the "Iceberg" effect can be found in many other statistics originating with the police; for example a much higher percentage of murders are reported than minor assaults or burglaries. And certainly the numbers of drivers found guilty of speeding or drunken driving are related to the resources devoted to detecting these offences; a Home Office study has estimated that only about one drink-driving offence in 250 results in a conviction. Thus, the changes from one year to the next in the official statistics relating to drunken-driving convictions are much more likely to measure changes in enforcement than changes in the amount of drunken driving. Many studies have found a strong positive correlation between expenditure on the police and recorded crime. It is accepted by most criminologists that police crime statistics have a very tenuous connection with crime.

The instructions for categorizing road accident casualties in Britain contain no guidance about how to distinguish a slight injury, which should be recorded, from one that is real but so slight as to be not worth recording. The decision about what accidents to record is ultimately subjective. It is likely to vary from victim to victim and recorder to recorder, and to be influenced by the priority that individual police officers and police forces place on road safety relative to other demands on their limited resources. It will also vary with the number of staff available to record and process the information. Because the number of injuries increases as severity decreases, a small move of the recording threshold up or down is capable of producing a large change in the numbers officially injured. When large numbers of police are diverted to the pursuit of bombers or murderers, or are embroiled in industrial disputes, or occupied containing civil unrest, the reduced number available for recording injuries, especially minor ones, is likely to lead to a raising of the threshold at which an injury is deemed worthy of recording – thereby producing a "safety improvement".

In all countries that report road accident statistics, it is the police, not the health services, that collect the information. The rectangles of the Severity Iceberg represent the numbers of casualties officially reported by the British police for 1993. A Transport and Road Research Laboratory study comparing the police numbers with hospital records has found a 21 per cent under-recording of serious injuries and a 34 per cent under-recording of slight injuries. Hutchinson, in a review of all such studies he could find around the world, suggests that in many other countries the degree of under-reporting of non-fatal injuries is much greater. In Ohio one study estimated that as many as 45 per cent of injuries are not recorded by the police, another in the Netherlands estimated 55 per cent, another in Finland found 75 per cent of casualties needing medical attention were not recorded in the police statistics, and in Sweden, for cyclists, the figure unrecorded was 80 per cent.

The volume of the unrecorded bottom of the Iceberg – real injuries not considered worth reporting or recording – is likely to be very large. The application of existing definitions of injury is at the discretion of the police, and variable pressure on police resources renders the exercise of this discretion inevitably variable. What gets recorded is likely to be a mixture of what the policeman "damn pleases" and what circumstances permit. The deeper one goes below the surface of Figure 5.3 the fishier the numbers become.

London can serve to illustrate some of the limitations of these numbers for understanding accidents. In terms of its population size it bears comparison with many countries, being only a little smaller than Sweden and considerably larger than Denmark or Norway. Figure 5.4a shows the great variability of road accident fatality statistics from one year to the next for the 32 London boroughs. The average population of each borough is about 200,000; although fatality statistics may be accurately recorded, for cities or other jurisdictions this size or smaller, short time-series and trends in fatality data should be viewed with great suspicion because of the instability of the small numbers. By contrast Figure 5.4b shows that for the same boroughs the injury data from one year to the next correlate extremely highly. Figure 5.4c showing the strong correlation between the level of policing and the level of recorded accidents lends support to the speculation that the small year-to-year variability in recorded injuries is at least in part a function of the size and stability of the bureaucracy that records them. The extreme outlier in Figures 5.4b and 5.4c is the Borough of Westminster. It has a high number of injuries relative to its resident population because of the large daily influx of non-resident civil servants and other office workers who work there. It is heavily policed because of its large day-time population and to ensure the security of Parliament and the central government.

The relationship between road accident fatality rates (Fig. 5.4a) and road accident injury rates (Fig. 5.4b) for London Boroughs (per 100,000 population) for two successive years. Figure 5.4c the relationship between the number of police tours of duty performed by uniformed divisional constables on street duty in 1986 (per 100,000 population) and numbers of recorded injuries (per 100,000 population) for London Boroughs. Source: Adams 1988.

Regression-to-mean and accident migration

Most safety measures are designed and implemented to deal with a problem that has become manifest in accident statistics, and most claims for the

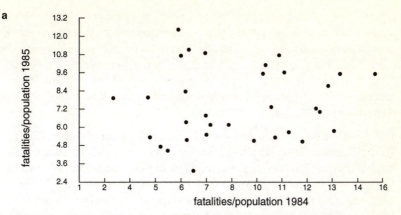

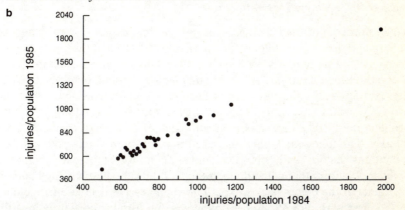

Figure 5.4 The relationship between road fatality rates (a, above) and injury rates (b, below) for London Boroughs (per 100,000 population) for two successive years. *Source:* London Research Centre.

(c) The relationship between the number of police tours of duty performed by uniformed divisional constables on street duty in 1986 (per 100,000 population) and numbers of recorded injuries (per 100,000 population). *Source of police data:* Metropolitan Police E Division.

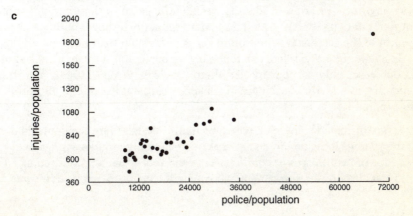

success of such measures rest upon a subsequent decrease in the statistics which prompted the measures. In addition to all the reasons discussed above for doubting such measures, there are two further sources of bias in the evidence upon which most claims for success rest. They are known as *the regression-to-mean effect* and *accident migration*. Ezra Hauer explains regression-to-mean as follows.

> Consider a group of 100 persons each throwing a fair die once. Select from the group those who have thrown a six. There might be some 16 such persons. (This is roughly analogous to the arranging of all road sections in order of increasing number of accidents and selecting the top 16 per cent). In an effort to cure the "proneness to throw sixes", each of the selected persons is administered a glass of water and asked to throw the die again. One can expect that all but two or three persons will have been cured. This "success" of the water cure is attributed entirely to the process of the selection for treatment.

The numbers of road accidents on any particular part of a road network bob up and down over time. After a particularly bad spell they usually come down. After a particularly good spell they usually go up. *Parts of the network that have experienced bad spells are defined as accident black spots.* When they are "treated", the numbers of accidents usually go down – but they probably would have gone down anyway. Hauer showed that ignoring this problem could bias the assessment of accident black-spot treatment enormously. The following table illustrates the magnitude of this potential bias. It contains "before" and "after" accident statistics for 2,637 road junctions in Sweden. It shows that *without any treatment* those junctions which had had accidents in the before period experienced an average decrease in accidents in the after period of 53 per cent, whereas one fifth of those junctions that had not had accidents in the before period had an accident in the after period. The effect diminishes with the length of the before and after periods, but the table shows that, even with a before period of four years, the effect is very strong. Table 5.1 is typical of the way data would be organized by highway engineers in order to identify *accident black spots* that required some form of safety treatment. And the accident reductions shown in the table are typical of the accident savings often claimed for such treatments.

Accident migration refers to a tendency for accidents at treated black spots to decrease, only to pop up elsewhere. Boyle & Wright (1984), note this tendency in a study of accident black spot treatment in London and proffer an explanation.

> . . . an untreated black spot, which by definition has more accidents than the average location on the network, also has a higher than average incidence of near misses or conflicts. This will mean that a proportion of drivers leaving an untreated black-spot will have been involved in some

form of conflict and will be driving more cautiously. Among commuters the effects of a near miss on one day may persist over a long period. This higher level of caution among a proportion of drivers will, it is argued, artificially deflate the numbers of accidents in the surrounding area. Successful treatment of the black spot will reduce the proportion of drivers leaving the black-spot who are behaving cautiously so that the number of accidents in the surrounding areas will tend to increase towards their "natural" levels.

Table 5.1 Regression to mean: injury accidents at 2,637 unsignalized and unaltered junctions in Sweden. [a]

Sites	Accidents per intersection during before period, 1972–5	Avg no. of accidents per intersection during equivalent after period, 1976–8	Difference (%)
2,039	0 (0)	0.19	+
441	1 (0.85)	0.42	−51
119	2 (1.70)	0.71	−59
24	3 (2.56)	1.33	−48
14	4.143 (3.53)[b]	1.50	−57

a. E. Hauer & B. Persaud, *Transportation Research Record*, 1983.
b. These sites had four or more injury accidents during 1972–5. Figures in parentheses are exposure adjusted figures to allow for greater length of the before period.

Boyle & Wright estimated that accident frequencies at treated black-spots *decreased* by 22.3 per cent but *increased* by 10 per cent in the immediately adjacent links and nodes of the network. They further noted that they had only attempted to measure a migration effect in the immediate vicinity of the treated black spot, but that the effect could well spread wider. This possibility points to yet another measurement problem; if the effect diminishes with distance from the treated site, it becomes difficult to distinguish from the background statistical "noise", but it could be considerably larger than the 10 per cent found by Wright & Boyle.

Accident migration can be temporal as well as geographical. In Chapter 2 (Fig. 2.1) it was noted that the fatal road accident death rate for children in Britain has halved since the early 1920s. But scrutiny of the road accident death toll for the 15–19 age group reveals a more than fourfold *increase* over the same period. Thus, it appears that the impressive reductions in child fatalities achieved by greatly restricting children's independent mobility represent not lives saved but death deferred. As traffic has grown and the perceived danger increased, parents have responded by delaying the age at which they grant their children licences to cross the road, ride their bikes, or go to school on their own (Hillman et al. 1990). This has had the effect of delaying the educational experience of coping with traffic directly. And when, in their teens, they can no longer be held in check by their parents, they are confronted, ill prepared, with a much more dangerous world.

There is of course the possibility that regression-to-mean and accident black-spot migration are simply two different names for the same phenomenon – the bobbing up and down around their "mean" or "natural level" of accident numbers over space and time. Boyle & Wright postulate a behavioural explanation, while Hauer invokes an unresponsive die. Or, more likely, regression-to-mean and accident migration may be complementary processes; drivers may be reacting to changes in the road environment *and* the accident numbers may be bobbing up and down in a random fashion. The quality of the data does not permit the question to be resolved conclusively. The studies done so far assume that the accident data provide a true record of accident numbers, but it is likely that the recording process itself will be governed by the laws of probability; whether an accident close to the minimum severity threshold for recording becomes an accident statistic or not, is as difficult to predict as the toss of a coin. But the available studies do show convincingly that claims for the successful treatment of safety problems that rest on simple before-and-after accident counts will inevitably exaggerate the effect of the safety treatment.

Such before-and-after counts are the basis of almost all claims for the success of accident black-spot treatments and the claim by Britain's Department of Transport that AIP [Accident Investigation and Prevention] techniques are the most cost-effective proven way of reducing road accidents. In support of this claim it refers inquirers to its *Accident investigation manual* where one finds a lengthy list of studies reporting on the accident savings produced by "accident black-spot treatment". Although the manual was published in 1986, none of the studies purporting to demonstrate the effectiveness of black-spot treatments post-dates Hauer's 1981 paper calling attention to the significance of regression-to-mean effects, and Wright & Boyle's work on accident migration. The most recent study cited in the manual, and the study most relied upon in the manual as proof of the effectiveness of black-spot treatment was published in 1971 and is a particularly blatant case of selective use of evidence.

The evidence presented by the 1971 study was compiled from responses to a survey in which respondents were invited to send *only* examples of successful accident prevention schemes. The study says, quite openly, "Reports were not at this stage asked for of road safety schemes that had not proved successful . . . [the results reported] do not, of course, claim to represent a random sample, or even a representative selection". But by the time the evidence from this study has been summarized in the *Accident investigation manual*, it is claimed that it shows the "likely benefits in terms of accident savings from application of remedies". Not only does the 1971 study ignore the biassing effects of regression-to-mean and accident migration, it takes the cream off the top of this biassed evidence, and is then presented 20 years later by the Department of Transport as proof that the black-spot treatments that it is now using will reduce accidents.

Cultural filtering

Cultural filters, as we have seen in earlier chapters, select and construe evidence to support established biases. They are particularly effective in cases where the available evidence is contested, ambiguous or inconclusive, a description that covers most controversies about safety. There are some threats on which all cultures can agree; all drivers alive today, for example, slow down when they come to a sharp bend in the road. Some cultures will slow down more than others, but they are in general agreement about the nature of the risk. The less clear the evidence, the greater the scope for the operation of belief and assumption, which have their roots deep in previously filtered experience of the world. This filtering process operates both directly through our five senses, and indirectly through extensions to our perceptual apparatus in the form of stories, news reports, statistics and research. These extensions are in effect pre-filters which embody the biasses of those who collect and process the evidence and then pass it on.

These pre-filters are essential and unavoidable. No government or research institution can gather and analyze more than an infinitesimal fraction of all the evidence about the infinitude of risks that exist in the world, and beyond it – those who balk at the invoking of infinity should bear in mind the threat of meteorites. Time and money will be spent only on those threats that governments and researchers think are the most important. To characterize a piece of risk research as biassed in a particular way is not, therefore, a criticism; it is an attempt to describe its essence. Each individual will have his or her own unique set of biasses; cultural theory provides a typology for reducing this variety to manageable proportions. The first step in applying this typology to road safety research is to search for bias.

Some can be detected. Over many decades, research, policy, legislation, education and highway engineering have all focused strongly on the safety of people *in* vehicles, to the neglect of the welfare of vulnerable road users – those on foot or bicycle. The safety measures adopted have created vehicles that are safer to have crashes in, and road environments that are more forgiving of heedless driving. Measures adopted in the interests of the safety of pedestrians usually take the form of movement-restricting barriers that oblige people on foot to travel further through tunnels or over footbridges. Road safety education for children tells them nothing about their rights as road users; it is devoted exclusively to inculcating attitudes of deference to traffic from a very tender age. The Royal Society for the Prevention of Accidents encourages parents to keep their young children on reins, and an official government safety leaflet aimed at parents, distributed through the schools, now insists that children under the age of 12 should not be allowed out on the streets unsupervised by an adult. Other Department leaflets urge elderly pedestrians not to go out at night unless they are wearing reflective clothing. And the safety advice aimed at cyclists stresses the danger of cycling to the point that all but the heedless and foolhardy are likely to give it up.

British government statistics on road safety reveal a similar bias in favour of vehicle occupants. Many tables are published describing the "risk" to car occupants of death or injury per vehicle mile travelled. No such statistics are published for pedestrians because no regular surveys exist of pedestrian exposure to traffic. Although large sums of money are routinely spent surveying the travel behaviour of motorists in great detail, no surveys exist of the amount of time children spend in circumstances where they are exposed to the danger of traffic. A survey of road safety research in other countries reveals a similar bias in favour of vehicle occupants, particularly in the USA. Haight & Olsen (1981) note that the low priority accorded to the rights of children is betrayed by the language used to discuss the accidents in which they are involved. A category commonly used in the safety literature for the classification of children injured in road accidents is "dartouts". Haight & Olsen suggest that a more appropriate label might be "children".

Contemplation of the threat of traffic to children suggests two possible remedies: the threat might be withdrawn from the children, or children might be withdrawn from the threat. The first possibility is rarely considered in the safety literature. The central message for both parents and children is the *normality* of traffic danger and the importance of deferring to it. The message has changed little over the years. Occasionally the impact of traffic on children's independence and freedom is acknowledged, but only by way of pointing out the anachronistic nature of such concerns in the modern world. Addressing this problem in a transport policy paper in 1976 the British Department of Environment observed that

> The seeds of these accidents [to children in traffic] are sown in traditions of independence and freedom; sometimes also in thoughtlessness and lack of care. Parents and children alike need to be educated in the dangers and the means to reduce them; the government have launched a campaign to bring home this lesson for life.

If road safety policy has the twin aims of making motoring safer for motorists and getting everyone else safely out of the way, then there is little need for detailed research into the activities of those who are to be displaced, and indeed little is to be found. Conversely, one would expect research effort to be concentrated on safety measures that do not interfere with the smooth flow of traffic. For years, according to the cost–benefit analyses of Britain's Department of Transport, the principal benefit of the new roads it has built has been *time-savings for motorists*; the inconvenience and delay suffered by people on foot whose neighbourhoods were to be bisected with traffic was a cost whose existence is not acknowledged by the Department's method of cost–benefit analysis.

The priorities of road safety research are manifest in the concern expressed in both Britain and the USA in recent years about the danger posed to motorists by trees alongside roads, "street furniture" in the form of sign and lamp posts, and other solid roadside objects. It is a concern to which specialist conferences have been devoted. As a consequence some States in the USA appear about to pass legislation requiring the replacement of rigid roadside mail boxes with breakaway types which do less damage to errant cars. And in Birmingham (England), where pedestrian fatalities exceed by a wide margin those of vehicle occupants involved in single-vehicle crashes, a policy has been adopted of siting all new lighting columns at the backs of footpaths in order to reduce the risk of injury to people in vehicles that stray onto them.

Noise and bias

How then does one account for the biases described above? Figure 3.4 (the risk thermostat with cultural filters), suggests that everyone has two cultural filters: one for the rewards and one for the costs. The perceived rewards of the activity, in this case motoring, that creates the danger will influence the propensity to take risks. In this chapter we have focused on the direct risks of motoring posed by vehicles colliding with each other, or with people or objects outside vehicles. But, in all cases of risk assessment, responses to threats are influenced by views about the desirability of the activity that poses the threat.

One way in which motoring registers its presence is through the sound it makes. Figure 5.5a comes from a study of the reactions to traffic noise of 693 people living in Paris. The study is described as one of the most successful attempts ever made to correlate physical measures of noise with noise nuisance (Kryter 1970). For some years I used the Paris study as the basis of a student project in which one group of students, armed with sound level meters, and another group, armed with questionnaires, collected comparable evidence in London. The correlation reported in the Paris study was never improved upon. Some people were extremely sensitive to low sound levels, and others were unperturbed by high levels. One year the sound measurers returned from their data-collection work on the Archway Road in north London to report that the loudest noise they had recorded, by far, was in the Archway Tavern when the band was playing.

Noise is unwanted sound; one person's music is another's noise. This was further demonstrated by another noise-measuring project in which I participated. The first flight of Concorde into London's Heathrow Airport was highly publicized in advance. A group of environmentalists opposed to

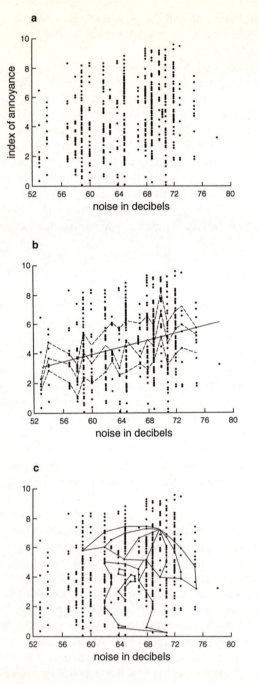

Figure 5.5 What noise annoys? *Source:* Adams 1977.

Concorde, equipped with sophisticated measuring equipment, positioned itself in Green Man Lane at the eastern end of Heathrow's southern runway to measure the "noise" made by Concorde's approach. Some Concorde enthusiasts had also gathered in the street to witness Concorde's first arrival. Concorde flew over and the sound meter registered significantly higher levels than the preceding subsonic jets. But the reading on the meter settled nothing. As Concorde passed over, several environmentalists clutched their ears in apparent genuine agony. At the same time the enthusiasts were leaping and cheering with a manic gleam in their eyes. What was being reacted to was not the intensity of the sound but its message. To one group (individualists?) it was the fanfare of progress; its note was the uplifting one of freedom and mobility and science in the service of man. To the other group (egalitarians?) it was the thunder of technology gone mindless; it was the ominous noise of the destruction of the ozone layer, and of civilization in retreat before the advancing multinational, offshore jet set. The frequencies and acoustic energies of both sounds were identical.

Figure 5.5b is characteristic of the hierarchist response to the noise problem – more earnest research. Noise has become a statistical term. "Noisy data" is information in which the relationship that one is searching for, and believes to exist, is masked by the influence of uncontrolled confounding variables. If a regression line is fitted to data that matches sound level exposure to "noise nuisance", it usually has an upward slope –consistent with the hypothesis that there is a positive correlation between sound level and nuisance. Reports of studies of this relationship commonly omit the statistical noise and present only the relationship depicted by the regression line. There are many confounding variables that a clever statistician might seek to eliminate or control for in order to reduce the scatter about the line of best fit in Figure 5.5b. If, for example, one earns one's living making cars or selling petrol, the sound of traffic will be the jingle of money in one's pocket. If one is a parent worried about the safety of one's children on the way home from school, traffic noise will have a threatening sound. If one is a young male motorcyclist a well tuned and amplified exhaust system will impress one's friends and terrify one's parents. If one is an environmentalist, traffic noise is the sound of pollution poisoning people and habitats being destroyed. And so on.

The problem is that attempts to generalize responses to "noise" by statistical methods such as regression will mask the underlying reality – "noise" is unwanted sound. It is the messenger and not the message.

At the time of writing, researchers commissioned by Britain's Department of Transport are continuing to try to refine measures of noise, and convert these measures into money so that noise can be incorporated into the cost–benefit analyses used to justify the Department's road-building projects. Figure 5.5c, a vole standing in the rain clutching its ears, is the response of a fatalist of my acquaintance who thought that Figure 5.5b looked like a join-

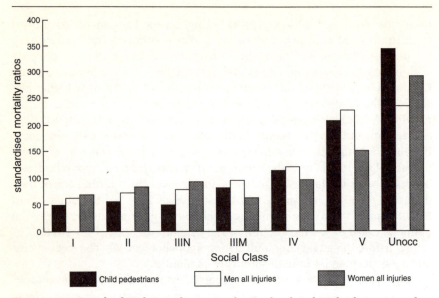

Figure 5.6 Standardized mortality ratios for England and Wales by age, gender, social class, and cause of death (*source:* Hillman et al. 1990).

up-the-dots game. Noise data and noisy data both offer scope for the exercise of bias. We hear, and see, only information that gets through our cultural filters.

The perception of, and concern about, risks is volatile. What gets through a cultural filter is often a matter of emphasis and presentation. Environmentalists, industry lobbyists and safety campaigners seeking to attract support for their causes, give tacit recognition to the existence of cultural filters when they debate amongst themselves the question "How can we get our message through to X?" – where X might be a civil servant, a politician of a particular persuasion, or a potential recruit to the cause in the form of the "typical" person in the street.

The finding of the Policy Studies Institute report, reported in Chapter 2, that children are being denied traditional freedoms by parents fearful for their safety, appears to have been screened out by the hierarchist filter of the Department of Transport – the official in charge of road safety dismissing such evidence as simply not relevant to his responsibility to reduce accidents. However, with the assistance of generous media coverage, it did manage to get through some individualist filters, and pricked the car-owning middle-class consciences of some parents, concerned that their children were being denied the freedoms that they had had as children.

Individualists are essentially egoists, but their egos can be stretched to embrace their children or kith and kin. Hence their susceptibility to evidence that their children are suffering from the growing threat of traffic. *Hierarchists* depend for their legitimacy on their ldefence of the entire hierarchy,

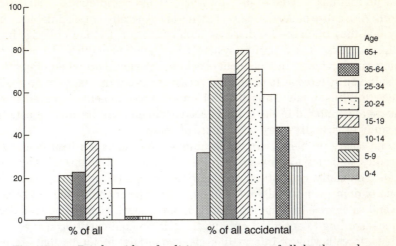

Figure 5.7   Road accident fatalities as a per cent of all deaths, and a per cent of all accidental deaths (Department of Transport 1992).

from top to bottom. Hence their susceptibility to charges that they are neglecting their duty of care for the lower orders. Both individualists and hierarchists, in the case of transport policy and road safety, are susceptible to worries that the benefits that they enjoy as motorists might be threatened if too many people become motorists. And *egalitarians* sometimes appear uncertain about whether they should oppose the car as a selfish and elitist form of transport, or seek to extend its benefits to everyone.

Fatalists are usually the outsiders in discussions such as this. They are the *lumpen proletariat* who do no research, who are sceptical of, or uninterested in, the research findings of others, and who are powerless to affect the course of their lives. They have poorly developed and unresponsive risk thermostats. They are close to the bottom of the socioeconomic heap, a position that suffers a disproportionate burden of risk, as Figure 5.6 indicates. They have low expectations of life, and have the lowest life expectancies in actuarial tables. In the young, especially young males, fatalistic alienation sometimes combines with vigorous individualism to produce a rampant nihilism, expressed on the road in the form of driving behaviour that can only be described as suicidal.

Off the road

Thus far in this chapter, the discussion of the problem of measuring risk has focused on road accident risks, for two reasons. Figure 5.7 shows that, in terms of lives lost, traffic is by far the largest cause of accidental death in Britain, as it is in most other motorized countries. But secondly, it is the cause of death that has – by far – the most accurate and comprehensive statistical record. As a cause of death, road accidents are generally not disputable in the way that they often are for diseases such as AIDS or deaths in old age from "natural causes". The recording system is independent of the health services, which can provide an independent check, and the numbers when aggregated are sufficiently large as to render the occasional recording lapse of no significance to annual totals. Thus, the problems of measurement discussed above that are encountered in the study of road accidents will be much greater in the study of virtually every other cause of death.

In all safety problems where the numbers of deaths are insufficient for mortality statistics to provide useful guidance, the problems arising from reliance on statistics relating to less serious incidents – injuries, morbidity, property damage, near misses – will be vastly greater. While the police collect and categorize information about non-fatal injuries caused by road accidents according to specified criteria, there is no equivalent agency for the systematic gathering of information about non-fatal incidents that are not related to traffic.

Insurance companies have the best available collection of information about incidents involving property damage, but the interpretation of their data, which for commercial reasons is rarely made available to outside researchers, also runs into the Severity Iceberg of Figure 5.3. Unknown numbers of incidents, on and off the road, are not registered for fear of legal liability, loss of no claims bonus, insufficient seriousness, or because the people to whom they happen do not have insurance. The premium income of insurance companies and the compensation payments made by them have experienced growth for many years. But this does not constitute evidence that the world has become more dangerous. It is evidence that more people can afford insurance, and is probably evidence of the growth of "moral hazard". *Moral hazard* is the insurance companies' term for risk compensation. It expresses their belief that people who are well insured compensate by being less careful; people who are insured against theft they suspect of being less scrupulous about locking up. Although the insurance companies apply a pejorative term to this behaviour, it is good for their business. It stimulates the demand for theft insurance; so long as premium income keeps pace, the greater the number of burglaries, the more profits those selling theft insurance will be likely to make.

Risk compensation can be observed in cases where insurance takes the form not of a contract with an insurance company but of physical protection such as, for example, a dam that controls flooding or reduces the damage it causes in the event of heavy rain. Dams can contain flood waters, up to a point. But if this insurance transforms a flood plain into land that is perceived as safely habitable, the occasional flood will be likely to cause much greater damage than that caused by the regular flooding of uninhabited land.

Countless other examples can be found of protective measures encouraging people to venture into danger. Rock climbers will attempt manoeuvres with safety ropes that they would not attempt without, similarly trapeze artists with safety nets, or steeplejacks with safety harnesses.

No satisfactory way has been found to compare quantitatively the benefits of such added protection with the losses should it fail. Losses measured in dollars, or pounds, or yen increase with affluence. The wealthy lose more because they have more to lose. But, despite many decades of trying, economists are no closer than ever to resolving the "inter-personal comparison of utility problem". They suspect that a single extra dollar is worth more to a poor man than a rich man; but how much more they cannot say. They do not have a scale with which to measure the difference. They cannot use dollars, because it is the variability of the dollar's value that they are seeking to measure; and no one has yet seen a "util". Until this problem is resolved, we can have no meaningful way of aggregating the losses of the rich and poor.

Near misses

Even less satisfactory is the measurement of risk by the collection of information about *near-miss incidents*. This is commonly resorted to in situations where there are few, or no, accidents to support the researcher's conviction that safety measures are needed. This approach to risk management is almost always rendered useless by risk compensation. Consider a busy road junction. One can see a mixture of vehicles – from juggernaut lorries to bicycles – and a mixture of people – from the old and anxious to the young and heedless – all managing, somehow to find a way through to get where they want, usually without a policeman in sight. The danger is manifest. Braking a split second late, moving a steering wheel an inch or two to the left or right, misreading the intentions of the other road users, misjudging their speed, miscalculating the friction between tyre and road, a simple lapse of concentration, any one of these or many other possibilities could spell instant death. Yet an observer of the junction would have to watch a very long time before seeing someone killed. The near misses are legion. The greater their number, the higher is likely to be the perception of danger and the higher will be the levels of vigilance of the participants in the drama, and the more likely they will be to take avoiding action.

A near miss is another name for a *potential* hazard. No actual harm has been done. A miss is as good as a mile. Or is it? The separation of those near misses that should be systematically recorded in a safety study, from the infinitude of other potential hazards that life holds will inevitably be guided by bias. It is an exercise that will reveal more about the beliefs and values of those responsible for the research than about "objective" hazards.

Near misses in road safety research are defined by the value of a critical time to collision [TTC]. One study (Brown 1991) declares that the generally accepted upper bound for this measure is 1.5 seconds; that is, if no effective avoiding action is taken, and the participants in the potential collision continue on their established courses for another 1.5 seconds, there will be an accident. But Summala, in an experiment that presented drivers with an unexpected stimulus (the opening of a car door) and noted their reaction times, found an average response time of 2.5 seconds and a range from 1.5 and 4 seconds. Reaction time is crucially dependent on *expectancy*, and this is an essential component of a motorist's cultural filter. It is an essential component of the filter through which *all* risks must pass. And expectancies, or levels of vigilance, are even more difficult to measure than near misses.

For all non-fatal measures of risk there is a further intractable measurement problem, variously labelled shame, guilt, responsibility, liability, stupidity, or the *Hawthorne effect*. Observing behaviour changes it; speeding motorists slow down when they see a police car, and video cameras in shops inhibit shoplifters. Accidents are unwanted and unintended events, and they are usually followed by an inquiry into who was to blame. There is a natural tendency for those who feel responsible to cover up. This tendency is sometimes amplified in the world of industry by financial incentives for safe conduct in the form of bonuses for accident-free or incident-free periods of work. The decision to collect data on accidents or incidents associated with a particular activity is a declaration that "the management" deems the safe conduct of that activity to be of particular importance. Hence the stigma attaching to a safety failure in that activity is likely to increase the under-estimation of failure by the officially recorded accident and incident statistics. *Stigma*, like risk, is a cultural artefact, and its power to distort accident and incident statistics will, as a consequence be variable.

Even death itself, the ultimate measure of accident loss, and a loss that is recorded in most countries with a high level of accuracy, is an unreliable measure of risk, because all deaths are not of equal magnitude. Look again at Figures 5.1 and 5.7. Accidents cause a tiny fraction of all deaths, yet accident prevention consumes a disproportionate share of attention and research effort. The number of newspaper column inches or minutes of television and radio air-time devoted to the 13,000 deaths by accident in 1991 greatly exceeded the coverage devoted to the 600,000 who died from all other causes. This suggests that we value some lives more highly than others. In some cultures life appears cheaper than in others. The death of a new born infant, a recent university graduate, a prosperous industrialist, a drunken tramp and a nonagenarian evoke different emotions in different people. The Treasury looks upon old age pensioners as a burden. The very old are cheered to the finishing line. Usually the losses of lives that held out the most "promise" are mourned the most. But there is no agreed metric by which such losses can be compared.

I have in this chapter skimmed the surface of the available evidence that casts doubt on our ability to measure risk objectively. Is there then no evidence on which all cultures can agree? Are there no safety interventions in which we can feel confident? Are there no safety interventions whose efficacy can be demonstrated objectively? In Chapter 7 I will examine a possible candidate, seat belt legislation. Hundreds of millions of motorists around the world are now compelled by law to belt up. It is a law that enjoys the support of a law-makers' consensus around the world. But first I examine a further measurement problem – the translation of risk into cash.

Chapter 6

MONETIZING RISK[1]

The risk thermostat model (Fig. 2.2) has at its centre a box labelled *balancing behaviour*. The *reward* of, say, getting to the church on time might induce a prospective bridegroom to drive faster and more recklessly than normally. In the terminology of our model this behaviour is accounted for by the driver balancing a higher than normal *propensity to take risks* with a higher than normal *perceived danger*. This propensity and perception are states of mind that are not directly measurable and are assumed to be responses to external conditions that have passed through cultural filters. The mental mechanisms by which such balancing acts are performed are but dimly understood, but behaviour is assumed to seek an "optimal" trade-off between the benefits of risk-taking and the costs.

However, any attempt to measure the costs and benefits associated with this behaviour would reveal that they are various, multifaceted and incommensurable. No one knows how the balancing trick is done, but certainly there is no evidence to suggest that the speed at which the bridegroom takes a corner is the result of a calculation in which either the benefits of getting to the church on time or the potential costs of a road accident are translated into cash. There is however a large body of literature in economics on *cost–benefit analysis* that insists that decisions can be *"rational"* only if they are the result of mathematical calculation in which all relevant considerations have been rendered commensurable; the common measure preferred by economists is money. The 1992 Royal Society report puts it this way:

> The optimum level of safety will be when risks have been reduced up to the point where the extra cost of any extra reduction just equals its benefits, but to go no further . . . . To weigh costs and benefits explicitly requires measuring them in common units and, so far, the only common unit suggested has been monetary value. (Marin 1992)

In *Blueprint for a Green economy*, popularly known as the Pearce Report

---

1. This chapter draws heavily on two articles by Adams: "The Emperor's old clothes: the curious comeback of cost–benefit analysis", in *Environmental Values* 2 (1993); and ". . . and how much for your grandmother?", in *Environment and Planning* A 6 (1974).

(Pearce et al. 1989), a book that has been influential in Britain in establishing the authority of economics in the current debate about environmental risks, the authors argue that "by trying to value environmental services we are forced into a rational decision-making frame of mind". In a later paper with Fankhauser on the greenhouse effect, Pearce (Fankhauser & Pearce 1993) again insists upon the reduction of all elements of the problem to cash as a necessary condition for rational decision-making.

A monetary assessment is crucial to design the optimal policy response. A comparison between the costs of greenhouse prevention and the benefits of avoided warming, which forms the backbone of an economically rational greenhouse response, is only feasible if damage can be expressed in monetary terms.

In *Blueprint for a Green economy* we find the following illustration of what this means when applied to risks:

Suppose that a particular programme involves a significant probability of a major catastrophe through soil contamination in a hundred years time. The cost of this contamination is estimated, in today's prices, to be £100 million and the probability that it would occur is 0.5. Then the expected cost in 2089 is £50 million. Discounted at 10 per cent per annum this amounts to £36, at 5 per cent it amounts to £3,802, and at 2 per cent it amounts to £69,016.

This is a straightforward application of the definition of risk most commonly found in the risk literature; probability is multiplied by magnitude, and, because the risk being discussed lies far in the future, the product is discounted to its *present value*. (The present value of some future loss is the economist's estimate of the sum that would have to be invested now at the going interest rate in order to produce a sum of money equal to the loss in the year in which it occurs.) In this example we find the Kelvinist view of risk carried to its logical conclusion; if a risk exists, it must exist in some quantity and can therefore be measured – and the only practicable measure, say the economists, is money.

In the discussion of "plural rationalities" in Chapter 3 it was argued that, in most disputes about risk in which the participants denounce their opponents as "irrational", or worse, the source of the problem is not a disagreement about the nature of rational thought, but the differences in the premises on which the contending parties have built their rational arguments. But in the use of cost–benefit analysis to make decisions about risk we encounter disputes about the nature of rationality itself. *Reason*, from the perspective of the economists quoted above, is reduced to *calculation*. In earlier chapters many grounds have been provided to doubt that numbers such as the one quoted above – a 0.5 probability of something happening in a hundred years' time – will ever be a sensible basis for deciding anything. But this chapter will focus on the proposition that decisions about risk will be improved by attempts to attach monetary values to such numbers.

Some problems

Cost–benefit analysts approach every project with the question "will it produce a *Pareto improvement*?" A Pareto improvement is a change that will make at least one person better off and no one worse off. Since projects that will produce this effect are rare, economists commonly modify the question to ask "will it produce a *potential* Pareto improvement?", meaning would it be possible for the winners to compensate the losers out of their winnings and still leave something over? In order to answer this second question, they need to have some way of comparing the winnings and losses, and this drives them to cash valuation. Because winnings and losses commonly come in a great variety of forms – ranging, for example, in the case of a transport project, from time-savings for motorists to the loss of life – it is necessary, the cost–benefit analyst argues, to make them all commensurable.

Until quite recently cost–benefit analysis was seen as a tool to be used in the context of partial equilibrium analysis for the evaluation of *projects*; its use was confined to relatively small schemes where the analysis could concentrate on localized effects. But in recent years the scope of its application has expanded enormously. In 1990 the first cost–benefit analysis of the greenhouse effect was conducted (see Ch. 9), and in 1991 Britain's Department of the Environment advocated expanding its use to the appraisal of *policy*. In *Policy appraisal and the environment: a guide for government departments* (Department of the Environment 1991) it urges policy-makers to reduce all the important elements of policy decisions to cash.

Where risk or uncertainty are encountered, the Guide recommends the use of *sensitivity testing*, *confidence intervals*, and the calculation of *expected values*. Sensitivity testing involves asking "experts" to judge the likelihood of the different possibilities, and then using these expert judgements in a "quantitative risk assessment" to identify the full range of costs and benefits and the likelihood that each will occur. It suggests using "the information about probabilities to derive ranges, or *confidence intervals*, to show how reliable any estimates are". The expected value of any outcome is derived in a way similar to the "expected cost", referred to in the italicized illustration from the Pearce Report on p. 94; it combines the probabilities and magnitudes of possible outcomes. The Guide provides a simple illustration: "under a policy there is a 10 per cent chance of the outcome being £10 million and a 90 per cent chance of it being £1 million, so the expected value is £1.9 million".

The Guide assumes that the necessary probabilities and monetized magnitudes will be available as required. It does concede that not everything relevant to policy-making can be translated *directly* into money, but it treats such factors as residuals whose values can be inferred from the values of those things that can be monetized. It tells policy-makers:

> The choices you recommend will imply a [cash] value for environmental resources in terms either of the other benefits forgone to preserve the environment, or of the other benefits gained at the expense of the environment . . .". (para. 4.8)

The Guide, like the Royal Society Report discussed in Chapter 2, anticipates resistance.

> The use of money as a standard is sometimes a barrier to wider acceptance. Most people believe that there are some things which are priceless (in the sense that they cannot conceive of any sensible trade-offs involving these things). It may be considered immoral to place a value on goods such as clean air and water which are generally seen as a right for all. *But* a monetary standard is a convenient means of expressing the relative values which society places on different uses of resources. Valuation is, *therefore*, a means of measuring public preferences, for example, for cleaner air or water, and is not a valuation of those resources in themselves (para. 4.15, my italics).

The Department of the Environment appears to be saying that the use of a money standard may not be sensible or moral, but it is convenient; *therefore* a money standard can be employed so long as you are only using it to measure *preferences* or *relative* values and not *actual* values. Perhaps this makes sense to an economist, but other readers are likely to need help.
The Guide tries to be helpful but only succeeds in deepening the mystery.

> Monetary evaluation is about measuring preferences. It is not about measuring intrinsic values of the environment (that is, values which some people may argue reside in the environment itself, independently of any human perceptions). Economic values and intrinsic values are different. Values in things are not measurable, though they could be taken into account in decision-making. (para. C1)

What meaning a human policy-adviser should attach to a value that is independent of human perceptions is not clear. Is it, for example, the value that a tree places upon itself? The Guide does not tell us, nor does it elaborate on the way in which decision-makers might "take into account" such values. The very existence of such values is left in doubt – "some people may argue" that they exist. Beyond mentioning their possible existence, the Guide has nothing further to say about them. Their neglect in the Guide suggests that, if they do exist, they are of marginal importance in comparison with economic values which can be expressed in monetary units.

David Pearce, the Department's economic adviser at the time of the Guide's publication, in a recent article entitled "Green economics", has another go at explaining economic value (Pearce 1992). He complains about his critics who have not taken "the trouble to investigate the meaning of economic value". He states

There is, of course, the view that "we cannot value the environment". But the meaning of this objection is not always clear, and confusion has arisen because economists have themselves used slipshod language. What economic valuation does is to measure *human preferences* for or against changes in the state of environments. It does not "value the environment". Indeed, it is not clear exactly what "valuing the environment" would mean.

Pearce (1992) cites Adams (1990) as "an illustration of the muddle that non-economists get into on economic valuation". I remain "muddled". Defining "economic values" as "preferences" does not get around the problem that cost–benefit analysis assumes that all risks can be reduced to sums of cash. The insistence that economic valuations measure preference for change and not the value of the thing being changed is puzzling. Whether one speaks of preferences or values, economic valuation still requires the use of money as a standard. Compounding this puzzlement, the Guide, and Pearce, insist that economic values have three components (para. C5):

total economic value = user values + option values + existence values

An existence value, it explains, is the value that a person attaches to an "asset" that he knows about but may never actually see – but whose loss or damage he would nevertheless regret. The Grand Canyon, the Norfolk Broads, the Flow Country and endangered species are examples cited. In many cases, the Guide stresses, they are likely to be very important. This may be an example of the slipshod use of language by economists about which Pearce complains. Having insisted on the importance of the distinction between "preference for change" and "the valuation of resources in themselves", and having said that cost–benefit analysis is concerned with the former and not the latter, the Guide proceeds to a discussion of the value of "assets", and speaks of clean air and peace and quiet being "traded in the property market".

Contingent valuation

Existence values, the Department states, should be measured by a process called "contingent valuation" – that is, asking people. In the case of potential environmental loss, the way they are asked is very important. They could, the Guide says, either be asked what they would be willing to pay to prevent the loss, or what they would be prepared to accept as fair compensation for the loss.

> Figures can be derived either for the willingness to pay (WTP) for an improvement (or to avoid damage), or for the willingness to accept (WTA) compensation for environmental deterioration (or to forgo environmental benefits). Studies show significant discrepancies that are still the subject of debate amongst experts. Most notably, values derived by WTP studies are often substantially less than values obtained by WTA measures. Where possible, attempts should be made to obtain both WTP and WTA measures, and to look for reasons for any divergence.

This advice blurs a long-established and important convention of cost–benefit analysis, namely that the potential benefits of a proposed investment project (or policy) should be valued in terms of what the beneficiaries would be willing to pay for them, and potential losses should be valued in terms of what the losers would be prepared to accept as compensation. The distinction made by cost–benefit analysts between *willingness-to-pay* and *willingness-to-accept* compensation is not an arbitrary convention; it is intended to ensure fairness. It acknowledges that the only acceptable judges of the value of losses arising from a project are the people suffering them. If, for example, a new road is proposed to go through someone's garden, how should the value of that garden be assessed? If the owner is unwilling to sell at the prevailing market price, that indicates that its value to him is greater than the equilibrium value established by willing buyers and sellers. The road builders could acquire the garden at its market price only by *compulsory purchase*. The difference between the compulsory purchase price, and what the owner would accept as fair compensation, is uncompensated loss, and, unless the value of this loss can be determined, it is not possible for the cost–benefit analyst to determine whether the benefits of a project exceeded its true costs.

In practice establishing WTA values for losses encounters an intractable problem. The only way to ascertain the value of something to an owner who is not contemplating selling is to ask him. No workable means has been devised for distinguishing honest answers from "bargaining" answers; some garden owners, aware of the strategic significance of their parcel of land, might hold out for a price higher than the one that would truly compensate them. But an even more difficult problem is created by other-worldly people who insist, honestly, that no amount of money would compensate them – this is an answer which, if taken seriously, the cost–benefit analyst can respond to only by entering an infinity sign in his spreadsheet. And it takes only one such number to blow up a whole cost–benefit analysis.

The Guide acknowledges that some studies have revealed "significant discrepancies" between willingness-to-pay values and willingness-to-accept values. This is an understatement. As Mishan (1971) points out, a willingness-to-pay value can be, literally, an infinitesimal fraction of a willingness-to-accept value, because the sum that an individual can pay for something (or to avoid something) is constrained by the limits of his budget, whereas the sum that someone might accept as compensation can be infinite. No amount of money, to use Mishan's example, is likely to compensate someone stricken with a fatal disease. Thus, the sum that a person is willing and able to pay to prevent a loss will rarely be an accurate measure of that loss to the person experiencing it. Pearce & Turner (1990) note that "economic theorists tend to dispute that WTP and WTA can differ so much simply because the theory says that they ought not to differ (and hence there must be something wrong with the empirical studies)". They do not say what the theory is that maintains that they ought to be the same, and most economists seem to accept Mishan's reason for expecting them to differ. Pearce & Turner acknowledge that psychologists "express little surprise that WTP and WTA are not the same". They conclude that "it seems fair to say that this problem is not resolved in the environmental economics literature" – and then pass onwards.

But the definition of "costs" and "benefits" is crucial to the choice of measure adopted. Table 6.1, based on an illustration originally used by Mishan (1971), shows the way in which the legal or moral context of a problem can transform a cost into a benefit. It represents the possible bargains that might be struck during a train journey by two travellers sharing a compartment – a non-smoker, and a smoker – depending on the rules of the railway company.

Under the *permissive rule*, which allows smoking, fresh air will be viewed by the non-smoker as a benefit – a departure from the status quo for which he expects to have to pay. The amount that he might pay will depend on the strength of his distaste for smoky air, and what he can afford. The amount that the smoker might accept to forgo his rights might depend on the strength of his addiction, his income, or his compassion – the exercise of which would produce "payment" in the form of moral satisfaction.

Table 6.1   Who pays whom?

	Smoker	Non-Smoker
Permissive rule	**Willingness to accept** compensation for forgoing the right to smoke	**Willingness to pay** for the benefits of a smoke-free journey
Restrictive rule	**Willingness to pay** for the right to smoke	**Willingness to accept** compensation for forgoing the right to fresh air

Under the *restrictive rule*, which forbids smoking without the agreement of fellow passengers, the smoker's willingness to pay will be influenced by his income and the strength of his addiction, and the non-smoker's willingness to accept will be influenced by his aversion to smoky air and how badly he needs the money. While it is difficult to imagine a civilized smoker requiring an extortionate sum of money to forgo his rights, it is possible to imagine a desperately ill asthmatic refusing a very large sum of money to maintain his air supply in a breathable state. In any event, only in exceptional circumstances are a person's WTA and WTP likely to be the same.

With respect to real-world environmental problems, one can find analogous situations. It does sometimes make sense to ask how much people might be prepared to pay to prevent certain environmental losses. The threat to Venice by the rising waters of the Mediterranean, or the threat of flooding in a river valley which could be protected by an upstream dam, are two examples. But these are both examples in which the "benefit" that people are being invited to pay for takes the form of preventing a loss that would otherwise be inescapable.

Most current environmental controversies, however, might be characterized as disputes between "developers" (representing the beneficiaries of a proposed project) and "environmentalists" (representing the losers), and the choice of which measure to use to value the prospective losses stemming from the project is, in effect, a choice of rule. If, in the above illustration, the smokers represent polluting industry, and the non-smokers the defenders of the environment, then to ask the environmentalists how much they are willing to pay to prevent damage to the environment is to assume a permissive law. It is tantamount to basing the cost–benefit analysis on a presumption in favour of "development". It is to assert that people have no *right* to clean air and water, to peace and quiet, to their architectural heritage, to cherished landscapes, or to habitats for endangered species. These are all transformed into privileges for which people are expected to pay out of limited budgets.

The Department of the Environment's Guide trivializes the difference between the two measures of environmental value. It acknowledges that WTP values for environmental losses are often substantially less than values obtained by WTA measures, but simply recommends trying to obtain both and "look for reasons for any divergence". It offers no further advice on which to choose, and in the section describing the procedure to follow for contingent valuation (p. 58) WTA is casually dropped from the discussion.

Perhaps a more significant clue to the bias of the government's appraisal of environmental losses is to be found in another British Government appraisal guide recommended by the Department of the Environment's guide. This is the Treasury's *Economic appraisal in central government* (Treasury 1991). Annex B of this guide contains the Treasury's recommended procedure for valuing "non-marketed outputs". It observes that

In addition to the intended consequences of a project, there will also usually be side effects. Thus, a proposal to build a new road will have an environmental impact along its route. Costs may arise as well as benefits. (p. 45)

Like the Department of the Environment's Guide, it concedes that not all these costs can be monetized, but goes on to argue that "non-marketed goods are generally best valued against the yardstick of an individual's *willingness to pay* for marginal changes in supply". The Treasury Guide contains no discussion of the problem of deciding between willingness-to-pay and willingness-to-accept measures. It contains no mention at all of willingness-to-accept measures. Its recommended valuation method embodies, therefore, a strong presumption in favour of development. It treats the entire population of the country as a group of non-smokers travelling in a smoking compartment.

There are obvious pragmatic reasons why the Department of the Environment and the Treasury should prefer WTP measures of environmental loss to WTA measures. As noted above, the DOE Guide accepts that, "most people believe that there are some things which are priceless". A person's loss cannot be separated from his belief about his loss. Values, like perceptions of risk, are inescapably subjective and cannot be detached from their valuers. If a person believes a loss to be beyond price, it *is* beyond price. And since "priceless", if it is to be included in a cost-benefit analysis, must be entered as an infinity sign, WTA values can seldom be used. They effectively give every loser a veto. There is no affordable test of whether or not a person's loss is genuinely beyond price. If WTA values are used, they must be arbitrarily limited by the economist to conform with his conception of "rationality".

It is a problem that has been acknowledged in the literature for many years. Pearce (1971) addressed the issue in his book *Cost–benefit analysis*.

One factor in the questionnaire [of the Roskill Inquiry into London's Third Airport] was the significant proportion of those interviewed who implied that no price would compensate them for movement away from their area. These replies would appear to be inconsistent with the general view that "each man has his price". If the response is ascribed to some element of irrationality in the householder, the problem arises of how to treat the factor in the cost–benefit analysis. The procedure in the study was to truncate the distribution at some arbitrary level. (p. 77)

Now, as then, the choice facing the economist is to truncate the distribution of reported willingness-to-accept values arbitrarily, or to use willingness-to-pay values when willingness-to-accept values ought to be used, or to abandon cost–benefit analysis as an unworkable method.

Death: the ultimate challenge

Not even life itself escapes the ambition of the serious cost–benefit analyst. The 1992 Royal Society report on risk contains an appendix entitled "Cost and benefits of risk reduction" (Marin 1992); it provides a clear and comprehensive summary of the state of the art of work in economics to ascertain the value of life. Marin insists that, in the evaluation of projects or policies that involve risk to life, "sensible decisions will involve monetary evaluation". He anticipates opposition from people who feel that "putting a price on life cheapens it in some underlying non-financial, but ethical, sense of value," and from others whom he characterizes as too squeamish to face up to harsh realities – people who "find it distressing to think about distasteful decisions involving the choice of life or death". But economists do not flinch from such decisions; Marin reviews the evidence from a now extensive literature, and concludes that a *statistical* life is worth between £2 million and £3 million.

The adjective "statistical" softens the blow somewhat, but what does it mean? Policies and large projects frequently save or cost lives. The value of these lives poses the ultimate challenge to monetizing rationality. Whereas it might be possible to get away with dismissing as irrational people who declare that no amount of money would compensate them for the loss of their homes or gardens, it is rather more difficult to dismiss a person who argues that no amount of money would compensate him for the loss of his life. Apart from the cases of a few self-sacrificing souls who want to leave their dependants well looked after, the WTA value of a human life is infinity, and this makes projects that cost lives difficult to justify with cost–benefit analysis.

As we have seen in the example italicized on p. 94, the usual method employed to value uncertain future losses is to multiply their value by their probability. But if the major catastrophe envisaged is not soil contamination but loss of life, the method will not work. If one assumes that in the future, as now, most people will still insist that no amount of money will compensate them for the loss of their lives, no matter how large a discount rate one employs, the present value of the loss will still be infinite. Marin acknowledges the problem.

> At first sight it might seem that the standard cost–benefit procedure would be inapplicable to projects and regulations affecting mortality: the equivalent question would seem to be how much those whose lives [were to be lost] would pay to continue living, or equivalently how much compensation would they accept to allow their lives to be lost. The problem with using such questions is that it is likely that most people would pay all that they could raise to save their own or close relatives' lives, and that some would require infinite compensation for its loss. Even if some would accept very large but finite sums to leave to their heirs, even

a single answer of "infinity" would be enough to throw out a project involving extra fatalities, or to reject any loosening of safety standards whatsoever.

So the cost–benefit analyst changes the question; Marin continues:

> However, in practice these questions are not the relevant ones for the assessment of changing risks. In most projects it is not the case that we know which particular person will live or die as a result of any variation in the level of risk. Instead there are typically very many people who will each have a slightly higher or lower risk of dying. In this context the relevant question is how much people will pay for a very slight reduction in their chance of premature death or how much compensation they would require to accept a slightly higher risk, when the probability of death is still far below one. For such changes in risk, honest answers of "infinity" are virtually inconceivable.

So, uniquely for the loss of life, the cost–benefit analyst adopts a valuation rule that permits the probability of a loss to affect the value of the loss. After surveying the substantial literature on this subject, Marin arrives at a conclusion that can be reduced to a simple equation.

$$r = (1/\Delta p)\, X$$

Where r is the value of a given change in the level of risk, and Δp is the change in the size of the population over which one death is spread, and X assumes values in the range £2 million to £3 million.

Thus, each change in the risk of mortality of 1/10,000 would be valued at between £200 and £300, and where $p = 1$ (i.e. certain death for a specifiable individual) r would be between £2 million and £3 million – the value of a "statistical" life. If X were set at £2 million, Marin points out that this formula would lead to the conclusion that UK drivers would be prepared to pay £20 per year for a 10 per cent reduction in their risk of a fatal accident.

Marin concedes that this method breaks down as the value of p approaches 1. It works, he says, only in cases "where the risk involved is well under certainty" because "there are reasons to expect that as risk gets much larger, then the compensation an individual would require will grow more than proportionately". Marin further concedes that "there is likely to be some limiting level of risk that is the maximum an individual would ever voluntarily accept under normal circumstances . . . [although] at present there is probably not enough evidence to be categorical about suitable maximum risk levels". In other words, at an as yet unspecifiable high level of risk, the problem of infinity returns. Conversely, following this reasoning, as the population over which the risk of a death is spread increases, the amount of compensation an individual would require will decrease more than proportionately. Put another way, *the greater the ignorance about the identities of the lives at risk, the lower will be their value*; the more ignorant people are about the risks of a project, the more favourable its benefit–cost ratio will be.

Marin acknowledges other problems. To ascertain the value of risks to life, the cost–benefit analyst must still ask people, and he is still dependent on the honesty of the answers he is given. Some causes of death (Marin instances cancer) are more feared than others, such as road accidents; and, further, "people are still more worried about potential causes of death where they feel helpless to influence their own fate . . .". These problems lead him to suggest that the £2–3 million figures, which were based on studies considering fatal accidents at work and for drivers, should be treated as "very much a lower bound for risks involving these other factors". He offers no suggestions about where the upper bound might lie.

Another problem concerns the age of death: "as people get older, and there is an increase in their chances of mortality from causes other than the source of risk being valued, the willingness-to-pay for this risk [sic] would be expected to decrease". He alludes to evidence to support this view, but does not cite any. He nevertheless recommends reducing the value of life for the assessment of projects affecting the elderly; by how much he does not say.

Yet another problem concerns the value that people place on the lives of others. Here we get into problems similar to those discussed above with respect to existence value. He cites a study of non-fatal risk from pesticides, which found that people's willingness to pay to reduce others' risks aggregated to more than the valuations they placed on risks to themselves. Allowing people a monetizable interest in the fates of others creates a problem that is difficult to manage because of the indeterminate numbers of people who might be concerned about other people. This problem leads him to speculate about the motives of people who claim to value other people, and about the problem of double counting.

> It is not clear that other people's willingness-to-pay for the reduction in risk [to others] should be fully added on to the own willingness-to-pay. It depends on motivation. For example, if the motivation is pure altruism in the special sense of one person reflecting what the other person would want for themselves, then the amount they would pay for the other's safety should not be included. It would be double counting to add this onto the person's own valuation . . . . Conversely, if the motivation is because one relative or friend would miss the other's company in the event of death because of exposure to risk, then relatives' and friends' willingness-to-pay should be added on.

He concludes that "the issue is too new for evidence yet . . . but . . . at the very least it strengthens the case for not using too low a valuation". His suggestion that an individual's interest is (should be?) confined to relatives and friends, and his omission of any reference to existence value, suggest that he would not recommend monetizing the motives that, for example, lead some

people to support charities interested in the welfare of people who are neither relatives nor personal friends, nor even compatriots. This omission is
understandable, because, once such people are allowed into the frame, the
boundaries of the valuation exercise are in danger of being overrun.

One final problem that Marin addresses is that of *very small changes in
risk*. He rejects the proposal of the British Health and Safety Executive that
some risks (below 1 in a million per annum) are so small that they should
simply be ignored. He argues that, if very large numbers are affected by very
small risks, the aggregate value of these risks might still be significant. He
concludes that "even if each person's risk is already low, the benefit of further reductions in risk should still be valued in the same ways as before and
compared to the cost of achieving the further reduction".

I have already discussed in earlier chapters the impossibility of obtaining
the objective measures of risk necessary for such an exercise, but it is worth
contemplating here the difficulties that would be encountered in attempting
to value them if they could be obtained. Readers are invited to consider how
they would respond to the question "What amount of money would compensate you for the imposition of an additional risk of 10^{-6} per annum?" How
would *you* estimate such a sum? What would it mean? Readers are invited to
try these questions out on friends and acquaintances. My experience suggests that you will encounter many "don't knows", and sometimes "who
wants to know?" and "why?"

The economists who seek the value of a (statistical) life claim to be trying
to help to make decisions that will maximize *our* welfare. *They encourage us*
to entrust *them* – society's acknowledged expert advisers on the value of
things – with such decisions. But because the information they need to make
these decisions is subjective and locked up inside people's heads, the only
way they can get at it is by asking us – the people. So, to the extent that you,
and your friends and acquaintances, doubt your ability to provide meaningful answers to the question "What is the cash value of life, or of an increase
in risk of 1 in 1,000, or 1 in 1 million?", the economist's advice will be meaningless.

There is reason to suppose that such numbers are meaningless abstractions even for economists who have a professional interest in their being
meaningful. Consider again the italicized quotation on p. 94 taken from Blueprint for a Green Economy. It contains a set of present values for a hypothetical disaster 100 years hence. Did you notice that the discounted values given
in this example are wrong by two orders of magnitude? They should be multiplied by 100, giving present values of £3,600, £380,200 and £6,901,600 –
not £36, £3,802 and £69,016. It is possible that this is simply a series of typo-
graphical errors. Such error, however, is not a trivial problem in quantitative
treatments of risk; there are many cases in which people conjuring with
complex models and large numbers have managed to misplace their decimal
points.

But, given the importance attached to human error in the literature on risk and safety, let us consider for a moment the possible significance of this particular "accident". It occurs in a section of the Pearce Report that is explicitly devoted to comment on the power of compound interest to reduce large values in the future to insignificance in the present. This suggests that it might have been an accident in calculation rather than typography. But let us be charitable and assume that it is the result of mere slips of the keyboard that got past the proof reader. The ability of such large errors to escape detection through all the processes of checking raises questions about the *meaning* of the numbers. The calculation and interpretation of such numbers are, supposedly, part of the process by which *society* should perform its risk balancing act. The calculation of such numbers should become, it is proposed by the cost–benefit analysts, the means by which government policy-makers help society to articulate its perceptions of danger. The numbers are intended to be compared with future benefits, also monetized and discounted, to guide the making of decisions about environmental risks.

Marin, in his appendix to the Royal Society report, having surveyed the state of the art of research into the value of life, concludes that, despite all the "uncertainties" about the meaning of estimates of the cash value of risks to life, they still have a valuable rôle to play. The Royal Society's 1992 report on risk ends as follows:

> ... despite some of the remaining uncertainties indicated, it seems clear that it is possible to value the benefits of reductions in risk .... there are strong reasons to suggest that a value of £200–300 for each change in the risk of mortality of 1/10,000 would be a sensible minimum value.
>
> Expressing the same value in the conventional and more convenient way (although misleading if used carelessly), the value of statistical life to be used in the cost–benefit of risk changes would be £2–3 million.

Cultural filters

Debates about environmental risks frequently feature exchanges of insults in which the participants cast aspersions on the morality and rationality of their opponents. It has been argued in earlier chapters that we should look for the causes of such conflicts in the participants' underlying beliefs and convictions. So let us now put cost–benefit analysis under the lens of cultural theory. Who uses it, and why? What persuades some people that, despite all the problems discussed above, the £2–3 million price tag on a statistical life is useful information? What persuades others (including this author) that it is meaningless?

Let us begin with the *hierarchists*, because the projects and policies that are subjected to cost–benefit analysis are almost always their responsibility.

They take a managerial approach to problems, and incline to a view of nature that makes it manageable. To the extent that everything in the world can be made commensurable with everything else, their job of management is made easier. Experts can be employed to estimate probabilities, and surveys can be conducted to ascertain monetary values; agonizing judgement can be replaced by mechanical calculation. They incline to compromise and pragmatism. They concede that cost–benefit analysis is not (yet) perfect, but insist that it is nevertheless the best tool available. And where reality resists reduction to a common denominator, they will force it by, for example, using willingness-to-pay measures for potential losses, or truncating willingness-to-accept measures at some arbitrary level. They will sacrifice the Pareto improvement principle, from which cost–benefit analysis derives its moral legitimacy, to expediency – or "convenience" to use the language of the Department of the Environment's Guide.

The Guide states, hopefully, that its proposed method of appraisal "helps to gain public acceptance of the chosen policies". But in practice such methods are useful only for preaching to the converted; cost–benefit analysis is almost always used not to make decisions, but to justify decisions that have already been made. I can think of no examples of major disputes about threats to the environment where cost–benefit analysis has successfully converted the opposition. The Guide admits that cost–benefit analysis cannot do justice to the concerns of people – and the Guide concedes that it is most people – who think that some things are priceless. But a method that dismisses such people as irrational does not persuade them, it antagonizes them.

The process of "development", which the hierarchists manage, very frequently poses risks to the environment. By consistently measuring threats to the environment by asking people what they would be willing to pay to prevent them, rather than what they would be willing to accept as compensation, the hierarchists betray a clear pro-development bias. The "official opposition" to most projects and policies that pose risks to the environment is lead by *egalitarians*. There are cases in which, for tactical reasons, they have couched their arguments in the language of cost–benefit analysis, but in most cases they approach arguments so framed with great suspicion. (At public inquiries into road schemes in Britain, the Government has declared that it is its "policy" to rely on cost–benefit analysis to assess road schemes, in effect warning objectors that arguments not couched in the language of CBA will be disregarded.) Being in opposition and on the defensive, they do not feel the same need for a comprehensive decision-making framework. They usually find that the things they value most highly are undervalued or ignored completely by cost–benefit analysis. Nature, cherished landscapes, endangered species, peace and quiet, health, buildings with nostalgic associations, neighbourliness, community cohesion, and the welfare of distant descendants – all these are examples of things that they perceive as threat-

107

ened by the process of economic development, of which the contested project is usually a part.

Their reaction to risk is the opposite of that of the hierarchist. As we have seen above in the discussion of risk to life, the higher the level of ignorance about a risk, the lower will be the value assigned to it in a cost–benefit analysis. For believers in the precautionary principle, ignorance has the reverse effect. The greater the level of ignorance about the consequences of doing something, the stronger the argument for *not* doing it. The main concerns of the egalitarians relate to things that are not traded in markets, that are not conventionally valued in money terms, and which are therefore difficult or impossible to value for cost–benefit purposes. Further they are usually most concerned with potential *losses*, and these losses under the Pareto rules should be valued in terms of the compensation that people would be willing to accept for their loss. When asked what sum of money would compensate them for the loss of most of the things listed above, they commonly give answers that must be interpreted as infinity. This generates frustration and hostility in the hierarchists; they respond by casting doubt on the rationality of people who cannot, or will not, name their price. The egalitarians respond by questioning the morality of the hierarchists, and offering cost–benefit analysis as an example of their crass materialism.

Individualists take a totally pragmatic view of cost–benefit analysis. They are the NIMBYs (not in my back yard) of environmental disputes. They are interested in calculations that purport to be concerned with collective welfare only to the extent that they impinge on them as individuals, although if pressed they sometimes justify their single-minded pursuit of self interest by appealing to the operation of The Invisible Hand that ensures that the aggregate result of selfishness is maximization of the collective good. If they like the result of a cost–benefit analysis, they will approve the method that produced it. They will use it if it suits their purposes, and, because it is a highly malleable technique, it sometimes will. As usual the poor old *fatalists* are preoccupied with less theoretical concerns and have no view on the matter.

Kakadu National Park: an example

Consider the case of Kakadu National Park in Australia. The *individualists* are represented by a wealthy mining company who wished to exploit a site in the National Park which is sacred to the penniless Aborigines who live there – the *egalitarians* in this example. The *hierarchists* holding the ring are represented by the Australian Government. The issue brings together three different agendas. The interests of the mining company and the Aborigines are relatively obvious and uncomplicated. The Government, historically, has a record of treating mining companies rather well and Aborigines rather badly, but public sympathy for the Aborigines in recent years has grown stronger. The Government resorted to cost–benefit analysis.

The interests of the Aborigines present an interesting challenge for cost–benefit analysis. What cash value does one put on something that is sacred? A cost–benefit analysis that adhered to the Pareto rules would have to ask the Aborigines what sum they would be willing to accept for the loss of something they hold sacred, and that which is truly sacred is not for sale at any price. So playing the game by the proper rules would simply give the Aborigines a veto. There would be nothing further to discuss. But resorting to the usual trick of substituting willingness-to-pay would also not produce a satisfactory result. It would bring the method into disrepute by placing a derisory value on the interests of the Aborigines; again the result would be so one-sided that there would be nothing to discuss. Because the sum that the Aborigines are able to pay to defend Kakadu is pitifully small compared to the fortune at the disposal of the developers, doing the valuation on a willingness-to-pay basis would have produced a walkover for the developers.

In an attempt to salvage something from their cost–benefit method, the economists of the Australian Resource Assessment Commission conducted a survey in which they asked a random sample of 2,034 Australians how much they would be *willing to pay* to prevent the mining of Kakadu. The answers ranged from $52 to $128 per year. The cost–benefit analysts argued that, since this sort of money, if given by every Australian, would exceed the revenue from the mine, the proposed mine had failed its cost–benefit test and should not be permitted. The process of asking large numbers of people and processing the answers with a lot of other numbers would give the decision-making exercise, they hoped, the appearance of serious deliberation.

The mining company simply dismissed the results as "nonsensical" and "unscientific" (Adams 1991). The mining company had a point. The numbers yielded by surveys such as the Kakadu one are nonsense because they float free of any context that can give them meaning. If Kakadu had been placed in a list of all the endangered species in the world, and habitats, and cultures, and works of art, and historic buildings . . ., and if the people had been invited to say, item by item, how much they were prepared, and able, to pay each year for their preservation, *and* required to hand over the money, then the figures might have meant something – but for each item, including Kakadu, they would have been exceedingly small.

If the numbers had been too small to yield the desired result (it is almost always the case that the desired result is known before the exercise is begun), the cost–benefit analyst would have had another card to play. Anyone can claim to place an existence value on anything. Indeed, although they do not express it in money terms, a growing number of people all around the world, spurred on by news reports and television documentaries, are beginning to take an interest in the survival prospects of remote endangered species, such as some varieties of whale, that they have little prospect of ever seeing at first hand. So, if the willingness-to-pay values of all the people in Australia were not enough to save Kakadu, the cost–benefit analyst need only cast his net wider – to embrace the whole world if necessary. And if even that did not produce the desired result, the cost–benefit analyst would have a final card to play; he could change the question in his survey from willingness-to-pay to willingness-to-accept, and that would be bound to produce a result in favour of the Aborigines, because it would be bound to yield at least one infinity – a result he could achieve without leaving Kakadu.

Conflicts such as that over Kakadu illustrate why cost–benefit analysis is only ever likely to polarize disputes about risk. Where opposing factions place high values on their own threatened interests and low values on the opposition's potential losses, taking a numerical average of these numbers will settle nothing. The method is more likely to inflame both sides by its irrelevance.[1] For the egalitarian, asking people how much they would be prepared to pay to prevent a part of their birthright being taken away, or to avoid being poisoned, is a form of blackmail – like the probing of an extortionist trying to find out how much a supermarket owner might pay not to have the goods on his shelves poisoned. From this perspective, even substituting a willingness-to-accept value for a willingness-to-pay measure does not make the method moral. To ask the Aboriginal inhabitants of Kakadu what they would be willing to accept for something that their culture holds sacred would be to attempt to corrupt them. Many non-market goods, the most important non-market goods, are defiled by attempts to measure them with the measuring rod of money; from the egalitarian perspective, rape preceded by cash compensation willingly accepted is indistinguishable from prostitution.

Confronted with this argument the hierarchist (the Department of the Environment 1991b) replies

> As for the "corruption" of the Aborigines, I am afraid I have, however reluctantly, to adhere to my view that they have as much right as anybody else to exercise choice. If they believe they are better off accepting compensation, then they have to be allowed to make that choice even if they might subsequently regret it. It is rather akin to the necessity to allow a child to make mistakes.

The hierarchist asks "How much?" The egalitarian says "It is sacred". The hierarchist replies "How much?" Monetizing risk settles nothing.

---

1. The same problem arises with different income groups. Christopher Nash, David Pearce and John Stanley address this difficulty in "An evaluation of cost–benefit analysis criteria" (*Scottish Journal of Political Economy* June 1975.) They note that it is commonly assumed that "on average, different income groups have the same tastes, and that their different patterns of expenditure are explained solely by their income levels." This assumption is implicit in most practical cost–benefit analysis, but it is of course, highly suspect.

Who wants to monetize risk?

From the perspective of cultural theory, cost–benefit analysis is a collectivist's tool. Neither fatalists nor individualists see reasons for becoming embroiled in the long-running debate about the theory or practice of cost–benefit analysis. The perceived need to monetize risk arises from the pursuit of Pareto improvements – measures that will improve collective welfare. But cost–benefit analysis is, more specifically, a hierarchist's tool. Nature, according to the hierarchist's myth, requires managing. In *Blueprint for a Green economy*, Pearce et al. (1989) assert that "Preserving and improving the environment is never a free option; it costs money and uses up real resources".

But this is true only if preserving and improving the environment are seen as activities requiring active management. Clearly, it would be wasteful management to spend money on preserving or improving something if the costs of doing so were to be greater than the benefits; rational decision-making about the environment requires, therefore, that *all* the relevant costs and benefits be priced.

However, nature according to the egalitarian's paradigm is not to be commanded, but obeyed; preserving and improving the environment require not more human interference with nature, but less. From this perspective, global-scale environmental degradation is the result of careless and excessive consumption. There are two ways a fat person can lose weight. The hierarchist's way – health farms, exercise machines, liposuction – uses up real resources. The egalitarian's way – walking or cycling to work and eating less – *saves* real resources. The egalitarian's method of losing weight does not require cost–benefit analysis; he does not need to calculate the cash value of being slimmer and then work out whether or not he can afford it.

Chapter 7

ROAD SAFETY 1: SEAT BELTS

Letter to The Times, *13 July 1908, from Colonel Willoughby Verner:*
Dear Sir,
Before any of your readers may be induced to cut their hedges as
suggested by the secretary of the Motor Union they may like to know
my experience of having done so.

Four years ago I cut down the hedges and shrubs to a height of 4ft
for 30 yards back from the dangerous crossing in this hamlet. The
results were twofold: the following summer my garden was smoth-
ered with dust caused by fast-driven cars, and the average pace of
the passing cars was considerably increased. This was bad enough,
but when the culprits secured by the police pleaded that "it was perfectly safe to go fast" because "they could see well at the corner", I
realized that I had made a mistake. Since then I have let my hedges
and shrubs grow, and by planting roses and hops have raised a
screen 8ft to 10ft high, by which means the garden is sheltered to
some degree from the dust and the speed of many passing cars sen-
sibly diminished. For it is perfectly plain that there are many motorists who can only be induced to go at a reasonable speed at
crossroads by consideration for their own personal safety.

Hence the advantage to the public of automatically fostering this
spirit as I am now doing. To cut hedges is a direct encouragement to
reckless driving.
Your obedient servant,
Willoughby Verner

In England in 1992 and 1993 "joyriding" and "ram-raiding" – breaking and
entering by driving a solid car through a shop window – were much in the
news. At the same time there were many advertisements on television
demonstrating how airbags make high-speed crashes survivable. One adver-
tisement showed a driver deliberately crashing his car in order to inflate the
bag to provide a pillow on which to rest his sleepy head. The latest reported

craze for thrill-seeking young men is stealing cars with airbags and driving them into walls. The higher the speed of the crash, the greater the glory for the survivor.

From the earliest days of motoring, up to the present day, it has been obvious that people modify their behaviour in response to perceived changes in risks to their personal safety.This phenomenon, now widely known as *risk compensation*, seems to most people mere common sense. About the only area where it still meets resistance is in the work of people with a professional interest in safety. This resistance can be found at its strongest in the debate about seat belts. Seat belt legislation provides a classic example of the cultural construction of risk. The strength of convictions about what this legislation has achieved is remarkably independent of objective evidence.

Around the world hundreds of millions of motorists are now obliged by law to belt up. The seat belt law, with minor national variations, probably affects more people than any other single piece of safety legislation. The first seat belt law came into effect in the state of Victoria in Australia in 1970,[1] and by 1991 over 80 jurisdictions worldwide had laws compelling drivers and some passengers to wear seat belts (Evans 1991). It is now a "truth", *almost* universally acknowledged, that these laws have saved thousands of lives. It is a "fact" endlessly repeated, not only on television and in the popular press, but in the scientific literature. Seat belts feature routinely in discussions of safety as an example of a measure that yields enormous benefits for minimal cost. The "success" of seat belt legislation in saving many lives is frequently cited by advocates of other public health measures as an example of the way legislation and regulation can reduce risk.

In a British parliamentary debate about seat belts in 1979, William Rodgers, then Secretary of State for Transport, claimed "On the best available evidence of accidents in this country – evidence which has not been seriously contested – compulsion could save up to 1,000 lives and 10,000 injuries a year" (*Hansard* 22 March).

Although the magnitude of the savings attributed to seat belts around that time varied, the claims made in the scientific literature prior to the passage of the British seat belt law in 1981 were consistently large. A report by the Transport and Road Research Laboratory shortly before the parliamentary debate in 1979 concluded "seat belts reduce deaths of car occupants by at least 40 per cent" (Grime 1979). Hurst, also in 1979, more than doubled this estimate: "belt use reduces the chances of fatal injury by about 83 per cent for drivers and about 80 per cent for front seat passengers". The Royal Society for the Prevention of Accidents produced a campaign pamphlet (1981) which claimed that ". . . for belted occupants the deaths were reduced by 77 per cent in full frontal crashes and 91 per cent in roll-overs". The pamphlet

---

1. The Ivory Coast is reported to have passed a law earlier, but no analyses of the result are available.

concluded "no other single practical piece of legislation could achieve such dramatic savings of lives and serious injuries". In the 1981 parliamentary debates that preceded the passage of the law, the claim that 1,000 lives and 10,000 injuries a year would be saved was repeated frequently, although some influential supporters of the law advanced even larger claims; David Ennals (1981), a former Secretary of State for Health, informed Parliament that not wearing a belt increased sixfold a motorist's chances of being killed in an accident.

Britain and the USA were among the last of the world's highly motorized countries to implement seat belt laws. Most other countries had done so in the early and mid-1970s. In 1978 in the USA, frustrated seat belt campaigners were presenting similar claims for the life saving benefits of a seat belt law to a Congressional Inquiry (DOT 1978):

- Mandatory safety belt usage . . . [holds] the potential to save 89,000 lives on the highways over the next ten years.
- The potential for saving lives right now is tremendous with estimates ranging from 10,000 to 20,000 lives a year.
- French police have estimated that seat belts have reduced fatalities in France by 63 per cent.
- Two separate studies [in Sweden] . . . found that seat belts reduced fatalities and serious injuries by 50 to 70 per cent, minor injuries by 20 per cent.
- The [German] government estimates that 1,700 deaths and 30,000 injuries are prevented annually by the use of seat belts.
- Occupant restraints is the largest highway safety issue that we have ever had since the automobile came on the scene. It is more important than the safety aspects of the Interstate, more important than getting drunk drivers off the road. In my opinion, it is the number one issue, and I base that on the profound benefits that can be obtained from occupant restraint.

By the time of the vote in the UK Parliament in 1981, the seat belt law had acquired an impressive number of influential sponsors: the British Medical Association, the Royal Society for the Prevention of Accidents, the Royal College of Surgeons, the Royal College of Nursing, the Royal Scottish Automobile Club, the Society of Automotive Manufacturers and Traders, and the Automobile Association. In the House of Lords debate, Lord Avebury (11 June 1981) offered this list of sponsors as compelling evidence for legislation. "Why, after all," he asked, "would these institutions seek to mislead the public?"

The answer, it appears, is that they misled themselves. At this time none of these institutions appeared to be aware of risk compensation and the possibility that there might be a behavioural response to the compulsory wearing of seat belts. The possibility had not been investigated in any of the studies they cited. Their support for a law rested on two sorts of evidence:

the effect of seat belts in crashes, and the effect of legislation in Australia. Britain's Transport and Road Research Laboratory had published a review summarizing the available evidence (Grime 1979). It presented abundant evidence that the wearing of a seat belt improves a car occupant's chances of surviving a crash. But it contained a significant caveat; it said that "for direct evidence on death, however, it is necessary to rely on recent Australian data". The Laboratory's review did not mention the possibility of risk compensation. None of the prestigious institutions cited by Lord Avebury, and none of the countries that followed the lead of Victoria in passing a seat belt law, produced any compelling new evidence. The law's supporters all cited the original Australian evidence, or other people citing the Australian evidence, or other people citing other people, and so on.

There was other direct evidence of the effect of legislation that could have been consulted, but the Laboratory did not explain why it chose to ignore it. This other evidence did not support the claims made for the law and, as we shall see in a moment, Australia was a particularly unfortunate example on which to rest their case. By 1981 there was evidence available from 13 countries that had passed seat belt laws. Figure 7.1 compares their road accident records with those of a "no-law" group of four countries that had not at that time passed a law. Together these 17 countries constituted an impressive sample; they contained over 80 per cent of the world's car population. The bars on the "law" graph indicate the dates at which seat belt laws were implemented, beginning with Australia and ending with Denmark, West Germany and Switzerland in January 1976. Around this time all 17 countries, with the exception of Australia and Spain, experienced marked decreases in their road accident death tolls. *Collectively, the group of countries that had not passed seat belt laws experienced a greater decrease than the group that had passed laws.*

The decreases shown in Figure 7.1 occurred in the aftermath of the 1973/74 energy crisis when the whole world was anxious about the adequacy of energy supplies, and was being subjected to advice about the energy-saving benefits of light-footed driving. The country that experienced the greatest decrease in the mid-1970s was Denmark, *before* its law was passed. As can be seen in Figure 7.2, after its law road deaths increased slightly.

Australia – the case that provided the main justification for most of the world's seat belt laws – stands out as the country whose road death toll varied the least between 1970 and 1978. The analyses that led to the seat belt claims all assumed that the rising trend of the 1960s would have continued, but for the seat belt law. Figure 7.3 is typical of these analyses. But, as Figure 7.4 shows, Australia when compared with most other countries was exceptional in *not* enjoying a substantial decrease in road accident deaths in the 1970s. Figure 7.3 is interesting for another reason; it also contains the first suggestion that less careful driving by belted motorists might displace risks to other road users, mainly cyclists and pedestrians. Although the evidence

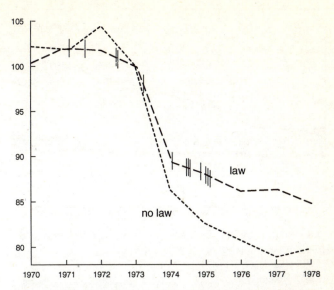

Figure 7.1 The effect of seat belt legislation. Indices of road accident deaths for countries with seat belt laws and without (1973 = 100). Bars indicate the dates at which laws came into effect in the "law" group (*source:* Adams 1982).

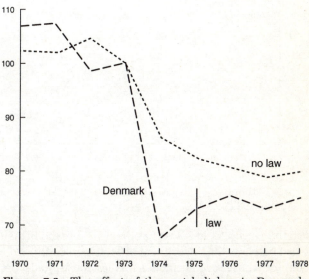

Figure 7.2 The effect of the seat belt law in Denmark (indices: 1973 = 100; *source:* Adams 1982).

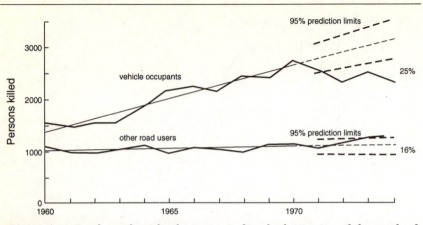

Figure 7.3 Road accident deaths in Australia; the beginning of the myth of seat belt effectiveness (*source:* Adams 1982).

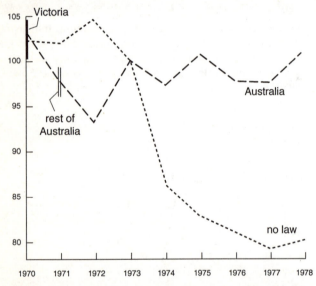

Figure 7.4 Australia's record compared to that of countries without seat belt laws (indices: 1973 = 100; *source:* Adams 1982).

summarized in Figures 7.1 to 7.4 was available before the British Parliament passed its seat belt law, the Department of Transport continued to insist that the only country whose road accident statistics constituted "direct evidence on death" was Australia, and that this evidence provided compelling support for a British seat belt law.

Now, over ten years later, with laws having been passed in over 80 juris-dictions, one would expect the evidence in support of the claims for seat belt legislation to be voluminous, but oddly it has shrunk dramatically. The claims now all rest on the experience of only one country, the United Kingdom. After surveying the global evidence, Evans (1991), in a comprehensive and widely acclaimed book on road safety, reaches the following conclusion:

> The highest precision evaluation is for the UK's law, where belt use rose rapidly from 40% to 90% in a large population of affected occupants. The law reduced fatalities to drivers and front-seat passengers by 20%. For smaller use rate increases, and for smaller populations (that is, in nearly all other cases), it is not possible to directly measure fatality changes. They can be reliably estimated using an equation based on the known when-used effectiveness of the belts together with a quantification of *selective recruitment* effects[1] – the tendency of those changing from non-use to use to be safer than average drivers. (p. 278).

In other words, out of the more than 80 jurisdictions with seat belt laws, only in the UK, according to Evans, was there a fatality-reduction effect that could be measured directly. In all the other jurisdictions the life-saving benefits were too small to register in the casualty statistics. (Evans does not name the exceptions to the "*nearly* all other cases" to which he refers, and with respect to the Australian claims he simply says "some estimates now seem to have been clearly too high"; he does not indicate what estimates he would now accept for Australia.) The claims made for seat belt laws in all these other jurisdictions rest on a deduction that *assumes* no risk compensation effect. Evans says "there is no evidence in the literature of measurable user responses to interventions that influence only the outcome of crashes, such as the use of safety belts or motorcycle helmets" (p.387). (In Chapter 8 I will discuss evidence from Evans's own research that undermines this contention.)

I do not dispute Evans's evidence concerning the life-saving benefits of seat belts *if one is in a crash*. The evidence that the use of a seat belt improves a car occupant's chances of surviving a crash is convincing. That a person travelling at speed inside a hard metal shell will stand a better chance of surviving a crash if he is restrained from rattling about inside the shell, or being ejected from it, is both intuitively obvious and supported by an impressive body of empirical evidence. Evans has calculated that wearing a belt reduces one's chances of being killed, if in a crash, by 41 per cent. He assumes that this benefit has been enjoyed by all those in the 80+ jurisdictions who belted up in response to a law, and the laws therefore can be given credit for saving

---

1. The evidence referred to by Evans concerning the "when-used effectiveness of belts" is based on crash testing using dummies, and on paired-comparison studies that examine the injuries suffered in crashes when one occupant was belted and another unbelted. "Selective recruitment effects" must be allowed for because the timid and cautious are most likely to belt up voluntarily, while the wild and reckless are most likely to defy a law.

many lives. But it does seem curious that with such a large effect, the only jurisdiction that he feels he can cite with confidence to demonstrate directly measured fatality reductions is the UK.

Doubt was first cast on the international evidence for seat belt laws in a paper of mine in 1981 (Adams 1981); Figures 7.1 to 7.4 above were first published in this paper. Britain's Department of Transport commissioned an internal critique of my paper. This critique, entitled *Seat belt savings: implications of European statistics* (Isles 1981), concluded that there was no foundation for the Department's oft-repeated claim that a seat belt law would save 1,000 lives and 10,000 injuries a year. It found what I had discovered, and what Evans found ten years later in his review of the evidence worldwide – that there were no directly measurable reductions in fatalities that could be attributed to seat belt laws. It said

> Available data for eight western European countries which introduced a seat belt law between 1973 and 1976 suggests that it *has not led to a detectable change in road deaths* [my emphasis] . . . The results are not compatible with the Department's "1,000 plus 10,000" estimates . . .

The author of this report was aware of the risk compensation hypothesis, and hence aware that evidence concerning the effectiveness of seat belts in crashes did not constitute satisfactory evidence about the likely effect of a law compelling people to belt up. He insisted that "international comparisons provide the only information about the effect of compulsory seat belt wearing, both on car occupants and on other road users". Furthermore, this report also noted that in all eight countries, as in Australia, the number of pedestrians injured following the passage of a seat belt law increased. Individually, none of the increases was statistically significant, but collectively this result was highly significant.

By the time the report was completed (it was dated 9 April 1981), the Department of Transport was already committed to a seat belt law. The report was suppressed and was not permitted to inform the Parliamentary debate which led, a few months later, to the passage of Britain's first seat belt law. The existence of the suppressed report was revealed by *New Scientist* almost four years later (7 February 1985). A leaked copy of the report has circulated widely since then, despite never having been published, and its conclusions were authoritatively, if belatedly, confirmed eight years later by an independent analysis by Janssen (1989) which concluded

> Time-series analysis was performed on car driver (and passenger) fatality rates for eight western European countries that passed seat belt legislation in the seventies. *There was no discernible effect of seat belt legislation on the fatality rate* [my emphasis].

Janssen (1991) also conducted what is at the time of writing the only experimental test of the hypothesis that seat belt wearing alters driving behaviour in realistic traffic conditions. He identified habitual wearers and non-wearers of belts and then, under the guise that they were participating in an experiment concerned with seat belt comfort, had them drive an instrumented car on a circuit including public highways and an off-road obstacle course. The habitual non-wearers drove the circuit belted and unbelted. None of the habitual wearers agreed to drive without a belt and Janssen did not attempt to persuade them. Janssen concluded that his experiment "yielded evidence both for selective recruitment and for adaptation effects in connection with seat belt wearing; [habitual] non-wearers drove faster than [habitual] wearers with the belt on; and [habitual] non-wearers showed a speed increase when they wore a belt". Other studies have attempted to measure differences in the driving behaviour of belted and unbelted motorists in traffic, but none of them constitutes a valid test of risk compensation. In none of these other studies were potential risk compensation effects separated from selective recruitment effects, with the result that they are all inconclusive – the risk compensation hypothesis suggests that people drive more dangerously when belted, while the selective recruitment hypothesis suggests that the safest drivers are the most likely to belt up voluntarily.

Thus, despite the fact that hundreds of millions of motorists all around the world are now compelled by law to wear seat belts, there has been only one small experiment (by Janssen 1991) to test whether or not their behaviour is altered as a result. The original Australian claims are no longer accepted, and only one jurisdiction, the United Kingdom, is considered by those who have examined the evidence to have produced an effect that is directly measurable in the accident statistics. Given the significance that is now attached to the United Kingdom result, I now look at it more closely.

The UK seat belt law

Unusually, as a concession to the doubts that had been raised at that time, Britain's first seat belt law was passed for a trial three-year period. It came into effect in January 1983, but was not made permanent until another vote in Parliament in January 1986. By this time the claim for lives saved had been reduced in a Department of Transport press release (15 October 1985) from 1,000 a year to 200. This figure was described as a "net" reduction; the decrease in the numbers of people killed in the front seats of cars and vans in 1983 was partially offset by an increase in the numbers of pedestrians, cyclists and rear seat passengers killed. This shift in fatalities was consistent with the risk compensation hypothesis that predicted that the added sense of security provided by belts would encourage more heedless driving, putting other road users at greater risk. But despite this implicit acknowledgement of risk compensation, the evidence on which Parliament relied when it confirmed the law in 1986 was fundamentally flawed. It ignored the effect of drunken driving.

Figure 7.5a shows what happened to road accident deaths in 1983, the first year of the law. Nothing remotely approaching the originally promised saving of 1,000 lives a year can be seen. There appears to have been a small, temporary drop below a well established downward trend. Most of the analyses presented to Parliament for the 1986 debate assumed that the slight upturn in the graph in 1982 represented a new upward trend, that would have continued into 1983, but for the beneficial effect of the seat belt law. The claims for the effect of the seat belt law are thus inflated by this assumed "ski-jump effect"; the actual fatalities were compared to the number expected on the assumption that 1982 represented the beginning of a new trend. However, it can be seen in Figures 7.5b and 7.5c that all of the increase in fatalities in 1982 was between the hours of 10 at night and 4 in the morning – the time known in the road safety literature as the "drink-drive hours". During the other hours the established downward trend continued. Figure 7.6 pinpoints the 1982 increase even more precisely; almost all of it occurred in non-built-up areas and was associated with drivers who had been drinking.

The decrease in fatalities in 1983 was clearly related to the campaign against drunken driving. In that year
- "evidential" breath testing was introduced
- unprecedented numbers of breath tests were administered
- the number of motorists successfully prosecuted for drunken driving increased by 31 per cent
- the decrease in road deaths between 10 at night and 4 in the morning was 23 per cent, while in all other hours it was only 3 per cent – in line with the prevailing trend
- the percentage of dead drivers who were over the legal alcohol limit dropped from 36 per cent to 31 per cent.

But the 1982 "alcohol blip" has never been satisfactorily explained. The sharp increase in that year in drink-related road accident deaths in non-built-up areas remains a mystery. According to a Transport and Road Research Laboratory Report (Broughton & Stark 1986) "the series for drinking car drivers in non-built-up areas shows an increase in 1982 which cannot be related to available explanatory variables".

In advocating the retention of the law in the Parliamentary debate in 1986, the Department of Transport relied most heavily on the analysis of two statistics professors, James Durbin and Andrew Harvey from the London School of Economics. The time-series models developed by Durbin & Harvey for their analysis of the seat belt effect were impressively sophisticated, but none of them contained alcohol-related variables. They attributed all of the decrease in fatalities in 1983 below the projected trend to the beneficial effect of the seat belt law, and none to the campaign against drunken driving. Durbin & Harvey presented their work to a Royal Statistical Society Seminar,

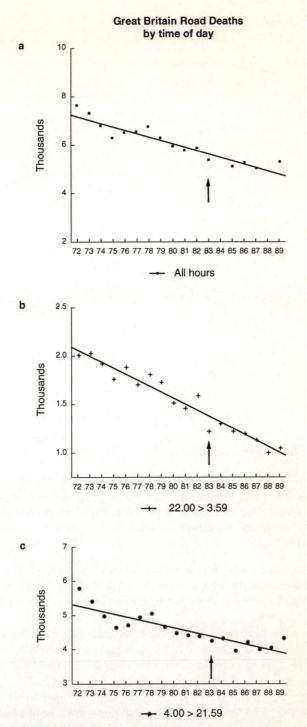

Figure 7.5 Great Britain road deaths by time of
day (*source: Road Accident Great Britain*, HMSO,
published annually).

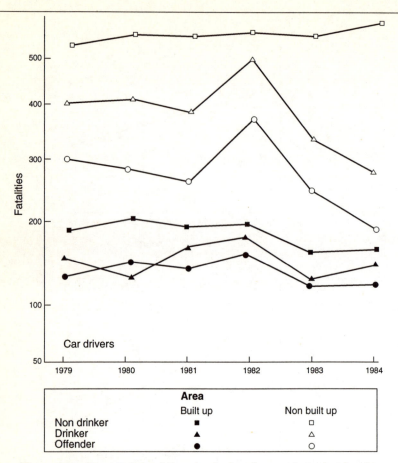

Figure 7.6 Great Britain driver deaths by place and alcohol level in dead driver. Both "drinkers" and "offenders" had alcohol in their blood; "offenders" were over the legal limited. Source: Broughton & Stark 1986.

and the discussion was published along with their paper. They acknowledged that their analysis had taken no account of alcohol and said "the study of the effects of alcohol is clearly an important area for future research" (Harvey & Durbin 1986).

But no studies have been done so far to explain why, after the seat belt law came into effect in Britain, seat belts have been so extraordinarily selective in saving the lives of only those who are over the alcohol limit and driving between 10 at night and 4 in the morning. It is a question that the Department of Transport has declined to pursue. A subsequent report from the Department on seat belts by Tunbridge (1990) still disregards the alcohol effect and claims all of the credit for the reduction in fatalities in 1983 for seat belts. It compounds this error of omission by disregarding the established downward trend of the data. It bases its conclusion, that the law saved lives, mainly on a comparison of data for three years before the seat belt law (1980, 1981, 1982) and two years after (1983, 1984). A glance at Figure 7.5 suggests that the established downward trends before 1983 continued. On the basis of these trends, one would have expected fewer fatalities in 1983 and 1984 regardless of any safety measures introduced in 1983. In the Tunbridge report this trend effect is also claimed for seat belts. (The probable causes of this trend are considered in Ch. 8.)

Further, the report cites Durbin & Harvey in a misleadingly selective way with respect to the effects of the seat belt law on pedestrians. Tunbridge says "they [Durbin & Harvey] concluded that there was no significant increase in the numbers killed and seriously injured subsequent to legislation". What Tunbridge fails to note is that the fatality statistics and the serious injury statistics tell different stories (see Fig. 5.2). The KSI (killed and seriously injured) statistical series is dominated by the much larger, but less reliable, injury numbers. Tunbridge does not cite the evidence from Durbin & Harvey with respect to the much more accurate fatality data on their own. Durbin & Harvey estimated that the increases in pedestrians and cyclists killed were 8 per cent and 13 per cent respectively. They also estimated an increase for rear seat passengers, to whom the law did not apply, of 27 per cent. Interestingly, the number of pedestrians and cyclists killed by heavy goods vehicles and public service vehicles (categories not covered by the seat belt law) *decreased* following the law. Using these categories as controls, the estimated increases in pedestrian and cyclist deaths following the law rise to 19.6 per cent and 40 per cent, although the small control numbers render these estimates unreliable.

The risk compensation hypothesis, and the historical time-series data on cyclist and pedestrian deaths, both suggest that the increase in cyclist and pedestrian fatalities following the seat belt law is likely to be a temporary transitional effect. Over the longer term, cyclists and pedestrians have responded, and are likely to continue to respond, to the increasing threat of motorized traffic by withdrawing from that threat (Adams 1988, 1988b, Hillman et al. 1990).

In summary there were two major road safety measures introduced by the British Government in 1983: the seat belt law and the campaign against drinking and driving. Figures 7.5 and 7.6 suggest that in 1983 there was a small, temporary, drop in road accident fatalities below the established trend. The evidence with respect to seat belts suggests that the law had no effect on total fatalities, but was associated with a redistribution of danger from car occupants to pedestrians and cyclists. The evidence with respect to alcohol suggests that the decrease in fatalities in 1983 during the drink–drive hours is accounted for partly by the still-unexplained rise above the trend in 1982, and partly by the drink–drive campaign in 1983. The evidence from Britain, which has been singled out as *the only jurisdiction in the world* in

125

which it is possible to measure fatality changes directly attributable to a seat belt law, suggests that the law produced no net saving of lives, but redistributed the burden of risk from those who were already the best protected inside vehicles to those who were the most vulnerable outside vehicles.

Three postscripts

1. In 1986, five years after the British seat belt law was passed by Parliament, risk compensation was effectively enshrined in the road traffic law of West Germany. Coaches fitted with seat belts now have a permitted top speed of 100 km/h, while those without are restricted to 80 km/h. The Royal Society for the Prevention of Accidents (ROSPA), a staunch believer in the life-saving benefits of the British seat belt law, when reporting this development in its journal *Care on the Road* (March 1987), unwittingly presented a classic example of the trade-off that routinely takes place between safety benefits and performance benefits:

   In Germany coaches with belts are allowed to travel faster than those without, thus allowing drivers to cover more miles in the hours they are allowed.

2. In September 1989, inspired by the "success" of the law compelling people to wear seat belts in the front seats of cars, the British Government made seat belt wearing compulsory for children under 14 years old in the rear seats of cars. Figure 7.7 shows the outcome measured in accident statistics. Comparing the year before (1988) with the year after (1990), there was an increase of almost 10 per cent in the numbers of children killed in rear seats, and of almost 12 per cent in the numbers injured; in both cases these increases were greater than the background increases.

3. On 10 November 1993 10 people were killed in a coach crash on the M2 motorway in Kent. Before any detailed information about how they were killed became available, there was an immediate chorus on television and radio, and in the press, demanding that seatbelts be made compulsory for coaches. The writers of editorials demanding seat belts in coaches, the authorities cited in the news reports, and the authors of letters to editors all had one thing in common: they rested their case on the "fact" that seat belt laws had saved enormous numbers of lives. It is highly unlikely that any of the "safety authorities" belonging to the chorus had any first-hand working knowledge of the statistical evidence. The vehemence with which they argued their case was rooted in a sincere belief in the efficacy of seat belt legislation. This belief is now so widespread, profoundly held, and insist-

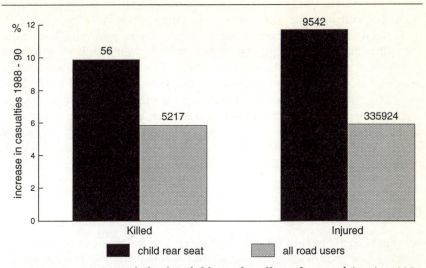

Figure 7.7 Rear seat belts for children; the effect of compulsion in 1989 (*source: Road Accidents Great Britain 1992*).

ently repeated that it is difficult to imagine any way in which it might be altered. The contrary view is routinely filtered out. I offer a personal example. On 11 November I had a long discussion with a journalist on a major national newspaper, and sent him a fax containing a summary of my view of the evidence which included Figures 7.5 and 7.6 above. His lengthy article appeared the following day. I had been edited out. He offered, when taxed with this omission, two not wholly consistent explanations. The first was that his article had been cut because of lack of space. The second was that no one else he had spoken to agreed with me. So powerful, and apparently unanimous, has been the demand in the popular media for seat belts to be fitted in coaches that it is difficult to believe that it will not soon be required by law.

The original claim that a seat belt law would save 1,000 lives a year in Britain was made at a time when there were about 200 billion kilometres travelled every year by unbelted motorists. The promise of the legislation was, in other words, that it would reduce the chance of death by 1 in 200 million per kilometre travelled. The change in behaviour required to offset the promised benefit would be equally small, and very difficult to measure directly – perhaps slightly faster or more aggressive driving, or the occasional extra lapse of concentration every few million kilometres. The promised reduction in risk was accompanied by a major advertising campaign to persuade people that a seat belt would make them very much safer.

Most readers of this book will now be habitual users of seat belts. You are

127

invited to consult your own experience. Might you drive a little bit more carefully if you were deprived of the protection of your seat belt? Some readers will be parents who insist on protecting their young children with rear seat belts or child restraints. Might you brake slightly more gently or corner a wee bit more slowly if your children were not safely secured in the back seat? The changes in your driving necessary to offset the loss of the protective benefits of seat belts are so small that accurate and reliable measurement by behavioural observation or introspection is extremely difficult. But all the world's seat belt laws assume that the answer to these questions is a unanimous "no".

Cultural theory

All this is fertile ground for cultural theory. The hundreds of thousands of people killed worldwide every year in road accidents constitute a large problem. Large problems and inconclusive information make people uncomfortable. The debate about seat belt laws appears to be as far as ever from resolution, despite more than two decades of experience and volumes of statistical evidence. Cultural theory suggests that in the face of such uncertainty the informational vacuum will be filled with contradictory certitudes; belief and conviction serve as substitutes for factual knowledge. Cultural theory further suggests that these beliefs and convictions assume distinctive and predictable forms, depending on their adherents.

Up until 1981 in Britain the seat belt debate in Parliament had become almost an annual event. It was a debate between the *defenders of life* and the *defenders of liberty*. The principal actors in these debates appear to have come from cultural theory's central casting department. The defenders of life were played with great conviction by *hierarchists*, and the defenders of liberty with equal passion by *individualists*.

The hierarchists believed that a law would save large numbers of lives and prevent even larger numbers of injuries. Their belief was conveniently rounded to 1,000 lives and 10,000 injuries a year. They argued that they had a duty to save people from themselves. If people were too ignorant, lazy or foolish to act in their own best interest, the state should intervene to ensure that they did. The hierarchist's research establishment produced abundant evidence to support the belief.

The individualists opposing legislation were heavily outnumbered. They had no supporting research organization and their campaigning support, such as it was, came largely from "right-wing" organizations such as the Institute for Economic Affairs. Their rhetoric was also consistent with the expectations of cultural theory. They were not opposed to the wearing of seat belts, they were opposed to the *compulsory* wearing of seat belts. They were supporters of law-and-order in so far as the law enforces contracts and protects private property, but they were opposed to criminalizing self-risk. Risk-taking is after all an essential ingredient of the private enterprise system in which they believe. Their epithet for a government seeking to impose a seat belt law was "the Nanny State".

They did not challenge the accuracy of the official estimates of 1,000 lives and 10,000 injuries a year to be saved, they dismissed them as irrelevant. They argued that, once the state began to criminalize self risk, there would be no logical stopping point; it would have to pass laws prohibiting rock climbing, cycling, drinking, smoking and eating too many cream buns.

Fatalists and egalitarians did not initially take much part in the debate. The fatalists never engage in debate about such matters because they cannot see the point. The egalitarians were indifferent because they did not see an issue that interested them. They had no reason at the time to question the validity of the claimed life-saving benefits of a seat belt law, and the law did not challenge the egalitarian way of life in any obvious way.

However, once risk compensation became an issue and the validity of the statistical basis of the benefits was challenged, they began to get involved. Once it became apparent that the law might shift the burden of risk from the rich and powerful and well protected –those in cars – to the poor and vulnerable – those on foot or bicycle – they began to see a cause worthy of their attention. Organizations such as Friends of the Earth, the Pedestrians Association and various cycling organizations began asking sceptical questions, and some came out in direct opposition to the law.

Their scepticism and opposition was rendered ineffectual by their late entry into the debate. It was also undermined by the nature of the evidence available to them. Like the individualists, they lacked research and publicity resources with which to counter the stream of confident assertion about the benefits of the seat belt law emanating from the hierarchy, a stream that continues to the present day.

As in many other cases the cause of the fatalists was appropriated by the egalitarians, who see it as their duty to defend the interests of the weak and vulnerable. The fatalists also received some tongue-in-cheek support from the individualists. Ronald Bell, a Conservative Member of Parliament well known for his right-wing views, was one of the most persistent campaigners against the law. Citing the precedent of an earlier law that exempted Sikhs wearing turbans from the requirement to wear motorcycle helmets, he proposed a religious amendment to the seat belt bill exempting "all Calvinists and other believers in predestination". The voting on the seat belt law found well known left-wingers such as Michael Foot (egalitarians) and right-wingers such as Enoch Powell (individualists) together in the opposition lobby.

Cultural filters

It has been argued in earlier chapters that we all perceive the costs and benefits of taking any particular risk through cultural filters. The Hans Christian Andersen fable of the emperor's new clothes is a good description of the way a cultural filter works. The Emperor and all his courtiers and all his subjects – with one exception – participated in a grand conspiracy of self-delusion. With the help of a pair of dubious tailors, they persuaded themselves of the exquisite textures and intricate patterns and beautiful colours and overall magnificence . . . of something that had no objective existence. The fable suggests that once an idea, however preposterous, becomes accepted by, and espoused by, established authority, it can become very difficult to dislodge. The idea becomes self-reinforcing. Authorities cite prior authorities, until the idea accumulates an authoritative pedigree. The idea acquires its own defence mechanism. Anyone incapable of seeing the Emperor's new clothes is "unfit for his station, or unpardonably stupid". The fact that large numbers of others believe the idea, can become sufficient reason for believing. After a while evidence is no longer required.

The evidence justifying the original seat belt law in Victoria initially seemed very convincing. Detailed studies of accidents, and experimental evidence with dummies, both supported the idea that *in an accident* a car occupant's chances of emerging unscathed would be dramatically improved by the wearing of a seat belt. The statistical evidence from Victoria, after its law was implemented, appeared to provide ample justification for compelling people to wear belts. The rising trend of 1960s in the numbers killed on the roads levelled off. The life-saving abilities of seat belt legislation became *fact*.

This fact became an integral part of the cultural filters of legislators all around the world, and became the basis of seat belt campaigns that culminated in the passage of laws in over 80 jurisdictions. In each country, a central plank in the case for a seat belt law was the list of other countries that had already passed one. As the list grew longer the plank grew stronger. The failure of the countries following Australia's lead to replicate its life-saving success did not appear to matter. Cultural filters become more efficient the more they are used, and the belief in the law was so deeply entrenched that the disappointing statistical results did not present a serious challenge. The "confounding variables" theory was invoked to explain away the results; one influential and exasperated researcher said "I just cannot accept that there is a sudden switch in driver behaviour just because the wearing of seat belts is made compulsory" (Mackay 1981). If one rules out the possibility of a behavioural change in response to the implementation of a safety measure, then any apparent lack of beneficial effect becomes evidence of the work of confounding variables; some other factor or factors *must* be at work masking the effect that you know to be there.

There is a parallel with the disputes discussed in Chapter 3 concerning the interpretations of dose–response relationships. It is now agreed by almost all road safety researchers that risk compensation occurs in situations where changes in the perceived threat are large (OECD 1990). In such situations the behavioural response to the threat is obvious and measurable. Such situations are analogous to "high-dose" experiments testing for carcinogenicity. But what of situations in which the change in level of risk is very low and difficult to perceive – in the case of seat belts, possibly of the order of 1 in 200 million kilometres travelled? As with low doses of toxins, there is a wide range of possible responses (see Fig. 3.6), all of which are impossible to measure in ways that will convince those not predisposed to agree.

Pressure to conform can be intense. Norman Fowler, the British Secretary of State for Transport at the time, was publicly accused at a British Medical Association conference of being "an accessory to mass murder" for his opposition to a seat belt law. My research casting doubt on the claims for seat belt legislation was denounced in a Parliamentary debate by a succession of MPs as "spurious", "eccentric", "preposterous" and "bogus". (See Adams 1985, Ch. 9; Davis 1993, Ch. 4 and Irwin 1985 also contain illuminating accounts of the conduct of the seat belt debate at this time.)

Research into the subject arouses strong emotions for entirely honourable reasons. Most researchers would claim to strive to establish the truth from a position of detachment above the fray. But if a researcher uncovers a truth with implications for the wellbeing of the public, he may see it as his duty as a citizen to pursue these implications into the realm of public policy. The fact that safety research involves matters of life and death creates a sense of urgency in both researchers and public policy campaigners. These are often one and the same, and as a consequence attitudes more appropriate to the latter sometimes intrude into the domain of the former.

Gatherings of road safety researchers tend to have an evangelical atmosphere. For example, in 1981, the American Association for Automotive Medicine and the Ontario Ministry of Transportation and Communications jointly sponsored a conference on seat belts (AAAM 1981). The papers presented to the conference were "scientific" papers devoted to examining the effectiveness of methods of restraint and the medical significance of such methods. But the purpose of the conference, summarized in the conference agenda, was to highlight "the need for physician commitment to influence public policy, research and education aimed at increased usage of occupant restraint devices". In 1984 the American Association for Automotive Medicine, this time jointly with the Society of Automotive Engineers, sponsored another scientific symposium entitled "Advances in Seat Belt Restraint Systems" (SAE 1984). The chairman's foreword declared the symposium's purpose to be "bringing recognition to the gravity of the crash injury problems and the benefits of seat belt use. A second foreword by Lee Iacocca, president of Chrysler, declared that "seat belts are the most effective device ever developed for saving lives and preventing injuries". And the leading scientific paper presented to the conference urged America to emulate Australia and make the wearing of belts compulsory, insisting that "a simple act of political courage would save countless American lives in road crashes". Although both conferences purported to consider scientific evidence about the efficacy of legislation, their real and publicly proclaimed purpose was to win adherents to their cause. Clearly, evidence that cast doubt upon the wisdom of this cause would have been unwelcome at these conferences. No such evidence was presented; it was filtered out before the conference.

The atmosphere in which the research is conducted can be morally intimidating. Anyone who cannot see the dramatic effects of road safety regulation risks being labelled by the *American Journal of Public Health* as an "ignorant nihilist" who is "symptomatic of a sick society" (Yankauer 1981). Lord Underhill in the House of Lords debate on seat belts (11 June 1981) declared "it would be terribly dangerous if credence were to be given to any arguments against the benefit of wearing seat belts". The danger that both Underhill and the editor of the *AJPH* feared was that if people had their faith in seat belts undermined they might stop wearing them. Truth, it has been said, is the first casualty of war; such is the passionate conviction of some safety campaigners that their campaigns can become crusades. In crusades heretics are not treated gently.

Such pressures can lead to a bias in the selection of the evidence that is published. Such a bias would seem to be the only way to reconcile the many small-scale studies, which appear to show a safety benefit following legislation, with the aggregate national statistics, which do not. For example, one much-cited study of casualty admissions to 16 hospitals in Sweden reported a decrease in admissions following the Swedish seat belt law of 29 per cent (Adams 1982). The fact that in Sweden as a whole after legislation the number of deaths and injuries to car occupants *increased* indicates that it must have been possible to find other sets of hospitals that showed an opposite result; but such a result has yet to be published.

Such selective pressures also appear to have been at work in the analyses of the effect of the British seat belt law that were presented to Parliament before it voted to confirm the law in 1986. Why was the effect of the campaign against drinking and driving omitted entirely from the analyses? Why was all the credit for the drop in road accident casualties attributed to seat belts? At least part of the explanation may lie in the fact that the seat belt law was on trial in a way that the drink–drive measures were not. The seat belt law had been passed for a trial three-year period and would have lapsed had it not been confirmed by Parliament. The proponents of seat belt legislation feared that an important life-saving measure might be lost.

Knowing that publication of findings that could undermine public confidence in a safety measure is likely to invite the charge of dangerous irresponsibility can lead to an editorial filtering of evidence. I have personal experience. In 1985 the Chief of Health Legislation for the World Health Organization commissioned me to write a review of the published analysis of the UK seat belt law for the WHO quarterly *International Digest of Health Legislation*. The contents of the review were a greatly abbreviated version of the story told above. The editor decided that it would be best if his readers remained in ignorance of the story. The review was rejected "for editorial reasons", reasons upon which the editor declined to elaborate further. The editor was concerned that his publication should not been seen to be associated with the review in any way. I was told that the WHO "would have no objection to the review being submitted by you for publication elsewhere, subject to the proviso that no mention is made of the fact that the review was commissioned and an honorarium paid by WHO".

Such pressure is also likely to encourage self-censorship by researchers in a way that can mislead. Convictions often masquerade as statistical hypotheses. If a research finding is consistent with the researcher's expectations and supports the (seat belt) campaign, he is likely to rush into print with the support of like-minded editors. Where a finding is contra-hypothesis, and would, if published, leave one open to charges of undermining public confidence in a measure believed to be effective, the researcher is likely to scratch his head and try again. The road safety literature is full of articles in which levels of statistical significance are dutifully reported. But what does it mean to say that some relationship is significant at say the 5 per cent level? It means that you might get a test statistic as large as the one you got by chance one time in twenty, even if there is no relationship between the variables tested. So if you sit enough monkeys in front of computer terminals working out correlation coefficients for sets of numbers taken from a random number table, and then publish the best 5 per cent, you are in danger of seeing significance where none exists. One in 20 is probably a very conservative estimate of the fraction of statistical tests done that are actually published; the mesh of the statistical/cultural filter through which most road safety studies are passed is almost certainly much finer.

Introspection

The above history of the seat belt debate is the version of one of the participants. More suspect still, it is a version from the losing side, or at least up until now. Cultural theory exempts no one from bias. The complete detachment of the hermit is an ideal state of mind to which a researcher aspires without hope of ever completely achieving it. Self-knowledge is more difficult than knowledge of others, or certainly feels so.

My earlier work on the seat belt issue preceded my acquaintance with cultural theory. Reviewing the seat belt debate for this book has stimulated much introspection, and contemplation of my own cultural filter. I conclude that my biases are context-dependent; *depending on circumstances*, I am capable of assuming all of the earthly personas of cultural theory. Sometimes I can see a rôle for government (hierarchist) action; I believe that it would be desirable to curb the depredations of the motor car. Sometimes, when confronted with extremes of power and vulnerability, I respond with an egalitarian's sense of injustice. Sometimes I am an individualist, resentful of the interference in my life of an overweening State bureaucracy. Sometimes, when contemplating the inevitability of my own mortality, I am a fatalist. As a dutiful researcher I strive for truth, detachment and objectivity, aware that I can never capture them.

Research, policy and action in the field of road safety are all hierarchist monopolies, or nearly so. Government and government-funded researchers decide which statistics to collect, and do most of the collecting and analysis. They filter most of the available evidence. They are predisposed to the view that it is possible and desirable to intervene in human affairs to reduce risk. They expect their interventions to work, and they believe their successes can be measured by accident statistics. Having examined the evidence from a position of as much detachment as I can muster, I believe the case for seat belt legislation to be fatally flawed. I see gross distortions in the evidence that has passed through an extraordinarily efficient hierarchist filter – so efficient that it has persuaded majorities in most of the world's legislative assemblies to pass seat belt laws.

Am I biased? Yes, inevitably. The seat belt law offends my individualist sensibilities; I see it as an unwarranted intrusion of state power into a realm that ought to remain the preserve of the individual. By shifting the burden of risk from those who are most powerful and best protected onto those who are weakest and most vulnerable, it offends my egalitarian instincts. But, perhaps most damning of all, it brings the hierarchy into disrepute. Far from curbing the depredations of the car, as its advocates maintain, it amplifies them. By its spectacular failure to deliver the safety benefits it promised, it has weakened hierarchy's authority to act in areas that are its proper domain. It must remain for the reader to decide whether my biases have undermined the validity of my argument.

Chapter 8

ROAD SAFETY 2:
MORE FILTERING

Safe roads and dangerous countries

All evidence bearing upon the problem of danger on the road, like all evidence bearing on all other risks, is highly filtered. For purposes of research and the design of safety measures, the filtering is commonly organized under three headings: the safety of roads, the safety of vehicles, and the safety of the behaviour of road users. On many occasions over the past 25 years I have had an opportunity to compare traffic in Britain with that in the USA and Canada. Under each of these headings the evidence that gets through my personal filter leads me to conclude that traffic in Britain is more dangerous; North American roads *feel* safer.

On arriving in the USA or Canada from Britain, one is impressed by the large size of North American cars, and by the fact that they appear to cruise around in slow motion on wide roads, with wide shoulders and grass margins beyond. Pedestrians in North America appear better behaved; they often wait at red lights even if there is no traffic coming, and there seems to be more help for school children in the form of warning signs and crossing guards. Traffic in residential areas also seems better managed, with many all-way stop intersections and street layouts discouraging through-traffic.

On returning to Britain, one is impressed by the much smaller size of the average vehicle, and the fact that it offers much less protection to its occupants in a crash. The roads are much narrower – most of them pre-date the invention of the car – and separation distances between oncoming vehicles are much smaller. Speed limits on motorways are higher (70 mph versus 55 mph), and enforcement is minimal – surveys suggest that over half the drivers exceed this limit. Traffic seems quick and darting compared to the more stately progression of North American traffic. Pedestrian behaviour seems anarchic; red lights are merely advisory for pedestrians, not mandatory. And the density of traffic is much higher in Britain; although Britain has fewer cars per head of population, it has more than twice as many per kilometre of road, and it provides, therefore, far more opportunities for vehicles to collide with each other or with pedestrians and cyclists.

By almost all the quantifiable criteria used by road safety planners and regulators to guide the design and implementation of their safety measures, North American roads should be safer: its vehicles are more crash-worthy; its roads are wider and straighter, its drivers drive more slowly and its pedestrians are more disciplined. And *subjectively*, British traffic feels to me, and to many others I have questioned over the years, more dangerous. But *statistically* the average North American is about twice as likely to be killed in a road accident as the average Briton (Table 8.1).

Table 8.1   Whose roads are safest?

	Road deaths in 1991	Road deaths per 100,000 population	Cars per km of road
USA	41,462	16.0	31
Great Britain	4,568	8.1	67

Sources: Transport statistics Great Britain 1993, and Road accidents Great Britain 1992.

This comparison between Britain and the USA is a national-scale example of the phenomenon introduced in Chapter 2 – the road considered dangerous by its residents, but safe by the statisticians who collect and analyze its road accident statistics. The fact that Britain appears to be twice as "safe" as the USA when measured by road accident fatality statistics is consistent with the difference between the two countries in their overall standardized mortality ratios for death by accident and violence illustrated by Figure 4.3. This suggests a large and persistent difference between the two countries in their collective propensity to take risks – on the road and off it; Americans appear to have their collective risk thermostat set higher. And this in turn suggests that the causes of this difference might be sought in differences in the two nations' risk cultures.

But this evidence is only suggestive. Risk, as we have noted earlier, comes in many incommensurable forms. Although the fatality statistics are the most accurate accident data available, they correlate very poorly, as Figure 5.2 reminds us, with the injury data; and almost no data exist to permit the study of their correlation with various forms of financial risk or social risk. What is clear, however, both from Figure 4.3 and from the comparison between Britain and the USA, is that success in reducing risk, if it is measured by fatal accidents (the preferred metric of most risk managers) has proved elusive. This elusiveness provides fertile ground for the operation of the contradictory certitudes described by the typology of cultural theory.

Safer vehicles?

The USA has led the world in most matters connected with the car; as recently as 1965 it contained over half the world's car population. It has developed a large bureaucracy to oversee the safety of its road transport system. Its National Highways and Traffic Administration is empowered to set vehicle safety regulations, and since the mid-1960s it has set many design standards requiring such things as padded instrument panels, tougher windscreens and energy-absorbing steering columns to improve the crash-worthiness of cars. Because the USA is the largest national market for cars in the world, these standards have been copied all around the world. All major car manufacturers now have departments whose job is to liaise and negotiate with governments over the drafting and implementation of these standards. Vehicle safety regulation has become a hierarchical affair involving large numbers of government officials, of grades high and low, and their counterparts in the automotive industry.

Perhaps the most influential British advocate of promoting safety through more crash-worthy vehicles is Murray Mackay; he contends (1982) that the US vehicle safety regulations have been a demonstrable success, and that considerable further progress is possible: "what the science of biomechanics has done is to show that the majority of deaths and injuries now occurring can be prevented by correct, crash-protective design". One of the most common ways in which those concerned with vehicle safety convince themselves that they are saving lives is by the presentation of statistics to show that over time there has been a large decrease in the number of deaths per vehicle or per vehicle kilometre (Adams 1987). Smeed (1949, 1968, 1972) demonstrated that large decreases in deaths per vehicle could be found in almost all countries as their levels of motorization increased. So common was it that he described the relationship between the level of motorization and the death rate per vehicle in the form of a mathematical "law".

Smeed derived his law from the relationship that he found between road accident fatalities and levels of motorization in 20 countries for the year 1938 (Fig. 8.1). He then looked to see how well this relationship applied to individual countries as their level of motorization increased. Figure 8.2 shows that Germany fairly consistently had more accidents than predicted and Great Britain fewer. The USA and the Netherlands conformed to the "law" rather better. But they all display a similar large decrease in deaths per vehicle as the number of vehicles increases.

Vehicle engineers have not been slow to claim credit for these large reductions. Increased traffic densities, they argue, have greatly increased the number of possibilities for crashes to occur, and modern cars with their better tyres, brakes and suspensions, their padded dashboards and laminated windscreens, and their crumple zones and safety cages, are manifestly safer than cars in the early decades of this century. It "stands to reason" therefore,

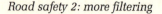

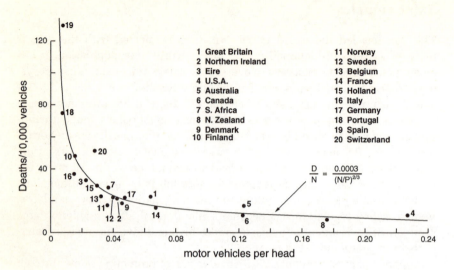

Figure 8.1 Relation between number of deaths per 10,000 registered motor vehicles and number of vehicles per person for 1938 (*source:* Smeed 1949).

that all the safety features engineered into modern cars deserve a large share of the credit for the reduced death rates. But do they?

Figure 8.3a applies Smeed's Law not to time-series data, but to fatality data for a cross section of 62 countries in 1980. The statistical fit first calculated by Smeed using data for a much smaller number countries 42 years earlier still fits remarkably well. Figure 8.3b, with the data transformed into logarithms, provides a clearer view of the residuals from the line of best fit. It shows the USA, with 70 vehicles per 100 population, at the bottom end of the line, and Liberia, with only one motor vehicle per 1,000 population at the top end of the line. Figure 8.3 also shows that Malaysia, with a death rate three times higher than that of the USA in 1980, has both a death rate and car ownership rate almost identical to those of the USA in 1925. Towards the top end of the curve one finds the world's poorest countries. They have very few vehicles, but the ones that they do have are lethal; if the Liberian death rate per vehicle were transposed to the vehicle population of the USA in 1980, the USA would have had over 6 million road accident fatalities, compared to the actual number, 51,000. The Third World countries towards the top end of the curve do not have their own car industries. All their cars are imported and have 80 years of safety technology built into them. Yet they are driven in such a way as to achieve kill rates per vehicle as high or higher than those achieved in the early decades of this century with Model-Ts and similar vehicles, which would be condemned by today's standards as outrageously unsafe.

Figure 8.3 suggests that the decrease in death rates per vehicle has much more to do with a behavioural response to the perceived increase in danger

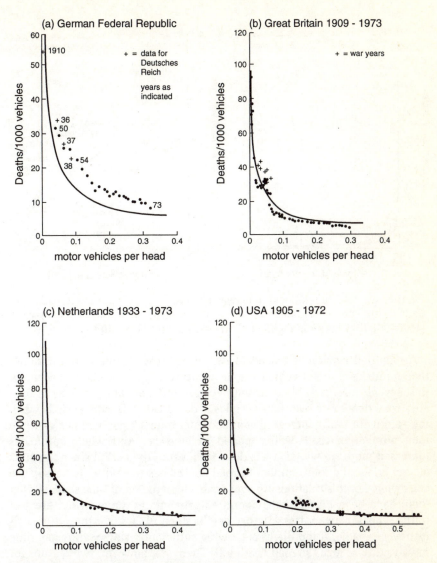

Figure 8.2 Relationships found by Smeed between road accident deaths per 10,000 motor vehicles (vertical axes) and motor vehicles per person (horizontal axes) for (a) German Federal Republic, (b) Great Britain, (c) Netherlands and (d) USA. *Source:* Adams 1985.

as traffic increases, than with the safety features built into modern vehicles. As Roald Dahl observed (Ch. 2), when the sight of a motor car is a rare event, children play heedless to danger in the street. When traffic is pervasive, they are not allowed out, and adults are more careful.

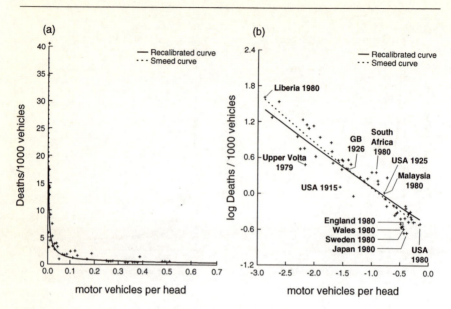

Figure 8.3   (a) The relationship between levels of motorization and road accident death rates per 1,000 vehicles of 62 countries of the most recent year
(1978–80). (b) Data from Figure 8.3a transformed into logarithms.

A modified version of Smeed's Law, which uses volume of traffic rather
than numbers of motor vehicles as the measure of exposure to road accident
risk, reveals a remarkably constant decrease (4.7 per cent per year) in Britain's road death rate over four decades (Adams 1987). During periods when
the volume of traffic increased at a rate faster than 4.7 per cent per year, the
total number of road deaths tended to increase. And when traffic has
increased more slowly, as it has done for most of the period since the 1973
energy crisis, the total number of road deaths has tended to decrease. The
consistency of this relationship over such a long period of time suggests that
there have been myriad behavioural adjustments to the growing threat of
traffic. The downward trend in road deaths since 1973 shows that we are collectively capable, if the threat is growing sufficiently slowly, of responding
in ways that reduce the road death toll. There are, however, two reasons for
not treating the downward trend in road accident fatalities as evidence that
Britain's roads have become safer. The history of death rates depicted by
Figure 4.3 suggests that fatalities tend to be displaced by safety measures into
other activities, rather than reduced. And, as discussed in Chapters 2 and 5,
accident statistics do not measure safety or danger; as traffic increases, the
death toll is contained, and sometimes reduced, by behaviour that avoids
danger rather than removing it. This can be seen most clearly in the restrictions now placed on children's freedom to protect them from traffic.

Safer roads?

The highway engineering equivalent to the manufacture of more crash-worthy cars is the construction of safer roads. Road engineers vie with vehicle engineers over which profession has contributed most to road safety; large claims are made for the safety benefits of road improvements. The most direct and specific of the highway engineering solutions advocated by road engineers is the set of measures commonly known as "black-spot treatment". The method is straightforward; it involves identifying those parts of the road network with exceptional numbers of accidents and subjecting them to whatever treatment is deemed appropriate by the road engineers.

We have already seen, in Chapter 5, that most claims for the effectiveness of this treatment are flawed by their failure to take account of the substantial biasing effects of regression-to-mean and accident migration. Accident migration may be, at least in part, an aspect of the regression-to-mean phenomenon. If accidents do tend to decrease at sites – treated or not – that have previously experienced high numbers, and increase elsewhere at sites that have previously experienced low numbers, it may simply reflect the probabilistic nature of accidents, or it may reflect a continuous behavioural response to perceived danger. Perhaps people respond to sites seen to be dangerous by being more careful – thereby bringing accident numbers down – and lower their guard elsewhere – with a consequent increase. In any event the likelihood that attempts to suppress undesired behaviour will lead to displacement rather than eradication is widely acknowledged in other areas of attempted social control, such as campaigns against crime and drugs. The phenomenon is well known to those at the sharp end of such campaigns – as illustrated by the following letter to local residents from the police chief in the part of north London where I live:

> Together I hope we can indeed persuade criminals to go beyond our boundaries in their pursuits so leaving Muswell Hill and Highgate free of all crime. (Inspector John Hare-Brown, newly appointed Unit Commander for Muswell Hill and Highgate in a letter to local residents associations, December 1987)

There is clear evidence that engineering improvements – whether to the car or the road – that increase a car's grip on the road will produce a behavioural response that offsets the potential safety benefit of the improvement. Figure 8.4 illustrates the results of a study (Rumar et al. 1976) that recorded the speeds of several thousand cars travelling around a bend in the road in Sweden. The vehicles were divided into two groups depending on whether or not they were fitted with studded tyres for winter driving. Speeds were recorded for high friction conditions (clear and dry) and low friction conditions (snow or ice). In high friction conditions there was no significant difference in the speeds of the two sets of cars. In icy conditions the cars fitted

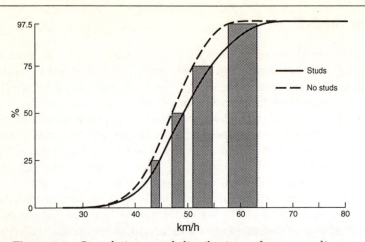

Figure 8.4 Cumulative speed distributions of cars rounding a curve in Sweden in low friction conditions (*source:* Rumar et al. 1976).

with studded tyres were driven faster than those fitted with unstudded tyres. The greatest difference in speed occurred at the highest speeds – at the speeds most likely to be associated with loss of control; the fastest two and half per cent of cars fitted with ordinary tyres were travelling at 57 kph or more, and the fastest two and half per cent with studded tyres were travelling at 63 kph or more.

On the question of whether cars fitted with studded tyres enjoyed a net safety benefit, despite being driven faster, the evidence was inconclusive – no accidents were observed. However, evidence from Ontario and further evidence from Sweden suggests that drivers respond to bad driving conditions in a way that decreases their chances of killing themselves while increasing their chances of having a minor accident. Figure 8.5 shows that over a ten year period in Ontario the number of fatalities was highest in August, when the roads are clear and dry, and lowest in February, when the roads are frequently covered with snow or ice. Injuries show a similar, although much less pronounced, seasonal variation. Figure 8.5 also shows that the severity of accidents, (measured by the proportion of injury accidents that are fatal), is highest in August and lowest in February. In Sweden where statistics are also published for damage-only accidents (Adams 1985) January and February are seen to have low numbers of fatalities but high numbers of damage-only accidents. Drivers appear to compensate for the hazardous conditions by slowing down, so that the accidents that they do have are much less serious. Traffic levels are also much lower when driving conditions are bad; people are advised not to drive unless their journey is absolutely necessary, and many appear to heed this advice. This response might be characterized as "gross risk compensation" – that is responding to a perceived threat by withdrawing from it completely.

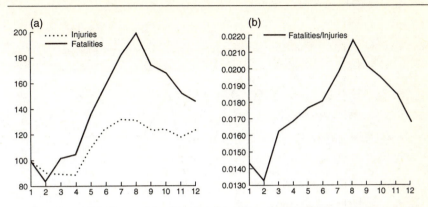

Figure 8.5 (a) Indices of fatalities and injuries in Ontario by month, 1970–9. The average for all Januaries (month 1) is set equal to 100. (*Source:* Adams 1985). (b) An index of road accident severity for Ontario by month, 1970–9. (*Source:* Adams 1985.)

Safer road users?

In Sweden in September 1967 all drivers were obliged to change from driving on the left to driving on the right. Overnight, people were obliged to reverse the training and acquired habits of a lifetime. Both motorists and pedestrians had to look left rather than right before crossing a road. In emergencies people had to jump or swerve in the opposite direction. Before the change-over there was widespread agonizing over the impending carnage. The result, as Figure 8.6 shows, was that Sweden had not had a "safer" September in many years. The danger associated with the change-over received enormous publicity. The perceived danger, it seems, was exaggerated, and people appear to have overcompensated. By October, people had begun to recover their nerve, and by November they were back to their normal (acceptable?) rate of killing each other.

A speculation

What would happen if safety regulations were to require all cars to be made of cardboard and fitted with inefficient brakes and a sharp spike in the centre of the steering column, if all roads were paved with a substance having the same coefficient of friction as ice (perhaps Teflon would do), and if all drivers were obliged to change sides every other month or, better yet, if there were no rules about which side of the road to drive on? The evidence surveyed above suggests that there would be no increase, and possibly a decrease, in road accident fatalities, but a substantial decrease in the efficiency of the

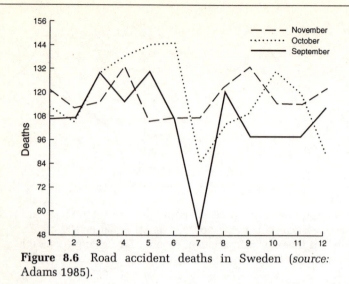

Figure 8.6 Road accident deaths in Sweden (*source:* Adams 1985).

road transport system. The potential safety benefit of most improvements to roads or vehicles is, it seems, consumed as a performance benefit; as a result of safety improvements it is now possible to travel farther and faster for approximately the same risk of being killed.

Bicycle helmets

At the time of writing the issue of bicycle helmets was the focus of media attention. It was receiving the same treatment that seat belts had received a few years earlier:

> To say they [bicycle helmets] encourage cyclists to take risks sounds like a re-run of the argument against car seat-belts. Editorial in *The Guardian*, 6 September 1993

"They said the same thing about seat belts" has become the most common way of dismissing doubts about the efficacy of other road safety measures, and indeed other non-road safety measures as well. The "success" of seat belt legislation forms an important element in the cultural filters of the proponents of crash protection and other forms of engineering and regulation that seek to protect people from the consequences of their own mistakes.

In Chapter 7 we saw just how efficient these filters can be. The fact that the numbers of children killed and injured in the back seats of cars *increased* after it became compulsory in Britain to belt them up, went completely unremarked in the road safety literature. The campaign that culminated in the children's rear seat belt law was considered a *success* by the campaigners; they got their law, and belt wearing rates by children in rear seats increased.

So firmly established is the "fact" of seat belt success that the casualty outcome was of little importance.

Road accident deaths and injuries increase and decrease over time in a way that is remarkably independent of safety measures. Whenever deaths and injuries do decline, the decline is seized upon as evidence that strengthens the initial belief in the latest measure; if they go up, they are either ignored or explained away by appeal to some variant of "the confounding variables hypothesis" – the most commonly invoked confounding variables are the economy or the weather. In complex systems such as transport in which vast numbers of factors are at work, confounding variables with which to construct a *post hoc* defence of a safety measure that fails to reduce casualties can almost always be found; but in the case of rear seat belts for children this has not, thus far, proved necessary. The perverse outcome has simply been ignored.

Opinion formers on newspapers, such as the author of the *Guardian* editorial quoted above, had their belief in the efficacy of seat belts increased by the "success" of the children's belt law. They also had their belief in the efficacy of measures *such as* the children's belt law strengthened. And when a subsequent measure, such as bicycle helmets, is called to their attention, they view it through an even more selective filter.

The case for bicycle helmets, like that of similar protective measures, appears compelling. If I bang my head walking through a low door, as I sometimes do, it hurts. If I were to do the same thing while running through the door, I might knock myself out. If I were to be riding through the door at speed on a bicycle, I might fracture my skull. If, in each case, I were wearing a bicycle helmet, the pain or injury would be diminished. There is a widespread view (shared by hierarchists and many egalitarians) that, if cyclists are not sensible enough to heed this argument and wear a helmet, they should be compelled to do so. In recent years the campaign for compulsion has been gathering momentum around the world. As with seat belts, the campaign began in Australia. At the time of writing, in Australia, New Zealand, and some jurisdictions in the USA, the wearing of cycle helmets is already required by law, and in Ontario a law has been passed that will take effect from October 1995. Such a law has a growing number of supporters in Britain.

In September 1993 a report was published by Mayer Hillman of London's Policy Studies Institute entitled *Cycle helmets – the case for and against*. It is the most comprehensive review of the literature on the effects of cycle helmets so far undertaken. It discloses a remarkable dearth of evidence that helmets are effective safety devices. Why, Hillman asked, should this be? He did not dispute that if you are going to bang your head you are better off wearing a helmet. But, for the first time in a major study of bicycle helmets, he also surveyed the growing body of evidence from other areas of risk-taking, that shows how the use of protective equipment and safety devices modifies behaviour. He observed that climbers without ropes, or cricketers and American football players without their helmets and pads, pursue their sports much more carefully without these safety aids. (One influence of helmets on the behaviour of American football players can now be seen on British television; the custom of congratulating a team mate for a good play by banging heads is not emulated by their bare-headed counterparts in British rugby.) Why, he asked, should cyclists not respond in the same way? He could find no evidence cited by the campaigners for a helmet law that addressed this question. He could find no evidence cited by the campaigners that acknowledged the existence of risk compensation. Hillman went on to chide the cycle helmet campaigners for a lack of consistency. He pointed out that *if* helmets were as effective in preventing head injury as they claimed, and *if* their aim was the prevention of head injuries, then they should begin their campaign with elderly pedestrians who account for more fatal head injuries than cyclists of all ages added together. Head injury fatality statistics, he continued, showed that children climbing and jumping, and motorists of all ages, also should be compelled to wear helmets before the law is applied to cyclists. Noting that there are three times more serious injuries to cyclists' arms and legs than to their heads, Hillman ironically advocated compulsory knee, elbow and shoulder pads as well. Pursuing Hillman's logic, there appears to be no obvious stopping point short of a world in which everyone is compelled to look like a Michelin man dressed as an American football player.

As with seat belts, Australia provided the first test of compulsion. Until then, all wearing rates everywhere remained low, and all the campaigns to encourage the use of helmets were based on evidence, not disputed, that helmets offered some protection from blows to the head. The main effect of the law in Australia, Hillman reports, was a reduction in the amount of cycling – a form of transport that he notes is health-promoting and environmentally friendly. The cost, inconvenience and discomfort of helmets all reduce the attractiveness of what would otherwise be a very cheap and spontaneously accessible form of transport; in Australia, after helmet use became compulsory, cycling decreased (according to the various partial surveys available) by between 15 per cent and 40 per cent. In an earlier study on the health effects of cycling for the British Medical Association, Hillman presented evidence that suggested that life-years gained through the health benefits of regular cycling, even in the present dangerous environment, far exceeded life-years lost in cycle accidents. In the Netherlands and Denmark, he noted, far more people cycle, few of them wear helmets, and fatality rates per kilometre cycled are between a quarter and a third of those in Britain. The way to make cycling safer, he concluded, is not to add protective armour to the cyclist, but to create an environment for cycling in which accidents are less likely to occur.

146

The reaction

Given its heretical message, Hillman's report was provided with a surprisingly high-profile launch. He was interviewed on *Today*, Britain's most popular morning radio programme, with an audience of over 5.5 million. Before the programme had finished, Peter Bottomley, a former Transport Minister who had been active in the promotion of cycle helmets, successfully insisted on being given air time to rebut a report that he had not read. Later the same day he reappeared on Radio 2 (another 1.5 million audience) to condemn the report's conclusion with ex-ministerial authority, and assert the benefits of cycle helmets. The Royal Society for the Prevention of Accidents joined in insisting that, although there might be something in the risk compensation hypothesis, it did not apply to the use of bicycle helmets. In terms of media coverage the report was a one-day wonder, and by the end of the day the report had been thoroughly denounced by "authorities" who had not read it.

I got a glimpse into the thinking of those responsible for the public presentation of the story when I attempted to interest a national newspaper in a review of the report. The editor responsible declined, worrying about what might happen if a cyclist were subsequently killed, having been influenced by an article in his newspaper not to wear a helmet. The *Guardian* editorial quoted above was typical of the media reaction. The evidence most commonly cited in defence of cycle helmets was the "success" of seat belts.

It is a commonplace that the print and electronic media are politically biased. People who do not share the bias complain of the slanted coverage of events. There is less agreement about the nature of the bias. The BBC frequently complains that it is assailed from all points of the political compass – because the bias hunters are themselves biased. While the issue of bicycle helmets is not overtly political, the above account of the treatment of the Hillman report provides another reminder of the inevitability of bias. Every day, newspapers, television and radio broadcast a combination of news and opinion. The intense competition between them, their appetite for controversy, and the frenetic pace at which they work leave little time for contemplation, reflection or independent consideration of new evidence. The result is a daily, or hourly, updating and reinforcing of established preconceptions.

Motorcycle helmets

Most highly motorized countries now have laws compelling motorcyclists to wear helmets. The issue appears to have been settled, and the safety campaigners have moved on to cycle helmets; indeed the evidence for the "success" of motorcycle helmet legislation is frequently used by campaigners for similar legislation for non-motorized cyclists. However, a return to the evidence that "settled" the issue reveals that its quality is comparable to that encountered in the debate about seat belts and cycle helmets.

An "experiment" was conducted in the USA in the 1970s that is widely believed to have proved conclusively that helmet legislation is an effective safety measure. It appears to have been an ideal "controlled" experiment. Over a few years a set of geographically diverse states, containing 47 per cent of the nation's motorcycles, repealed laws compelling motorcyclists to wear helmets. It was widely predicted that these repeals would cause a substantial increase in the number of motorcyclists killed. The motorcyclist fatality statistics appeared to confirm the predictions.

In the USA one of the most authoritative and influential publications on the subject was the report to Congress by the National Highway Safety Administration (NHTSA) in 1980. Figure 8.7 is a reproduction of the front cover of this report. The graph was taken from the body of the report and put on the cover because, presumably, it was considered the report's most compelling evidence in support of legislation. The report concluded that the decline in helmet use associated with helmet law repeal was "the single most significant factor" responsible for the dramatic increase in the motorcyclist death rate after 1975. This graph was considered so compelling that it was reproduced in another influential publication on the subject a year later – *The American Journal of Public Health* (Watson et al. 1981). The *British Medical Journal*, in an editorial entitled "A grim experiment" told its readers (9 August 1980):

> The immediate effect was a drop in the proportion of motorcyclists using helmets from 100% to 50%. Deaths from motorcycle accidents rose by an average of 38% in the states which had repealed their laws, while remaining constant in the other states . . . . Deaths and injuries on the road are one of the few subjects where preventive medicine can be based on reliable statistics on the effects of intervention . . . . The refusal by successive governments to take action on these data is a continuing disgrace.

But the NHTSA based its conclusion on data that were too highly aggregated to support any defensible inferences about the effect of helmet legislation. Figure 8.8 shows that, when disaggregated, the data do not support the NHTSA's conclusion. If the death rates for states that repealed their helmet laws and those that did not are plotted separately, after 1975 we can see that the blame for the increase in the death rate cannot be placed on helmet law repeal because, for most of this period, the increase was greater in the states that did not repeal their laws. In 1976 9 states repealed their helmet laws, in 1977 a further 14, and in 1978 a further 4. Between 1975 and 1978 the death rate in repeal states increased by 46.7 per cent while in the non-repeal states it increased by 48.2per cent. In 1978 the death rate decreased by 2.6 per cent and by 7.8 per cent in the non-repeal states. Throughout this period the death rate in the repeal states was, on average 19 per cent lower than in the non-repeal states (Adams 1983).

What did cause the sudden large increase in motorcyclist fatalities? Unfortunately satisfactory data do not exist to test a likely hypothesis. Figures 8.7 and 8.8 are graphs of fatalities per 10,000 motorcycles. The number of motorcycles is not a satisfactory measure of exposure to the risk of a motorcycle accident, because it does not allow for variation in the distance that the average motorcycle travels. In Britain where such data do exist, between 1973 and 1978 (in the aftermath of the energy crisis) the distance travelled annually by the average motorcycle increased by 44 per cent. If the distance travelled per motorcycle in the USA increased in a similar fashion over this period, and if the total distance travelled by motorcycles had been used as the exposure denominator instead of numbers of motorcycles, there would

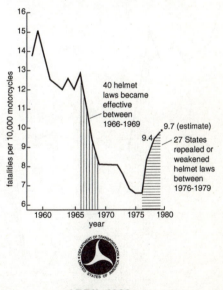

Figure 8.7 Front cover of NHTSA Report to the Congress (*source:* Adams 1985).

149

have been no large increase in the motorcyclist death rate per unit of exposure after 1975 (Adams 1983).

In Britain a law requiring the wearing of helmets by motorcyclists came into effect in June 1973. In 1973 fatality rates per 100 million miles travelled by motorcyclists decreased by 2 per cent; fatality rates for all classes of road user decreased by 9 per cent. In 1974, the first full year of the law's operation, fatality rates for motorcyclists increased by 2 per cent, while for all classes of road user they decreased by 3 per cent.

As with seat belts, helmets do offer protection in a crash, but there is no convincing evidence that they have reduced the numbers killed. Again, the most plausible explanation for the available evidence is risk compensation – a behavioural response to the added sense of security provided by the safety measure. For years I have been conducting an informal survey of motorcyclists. I ask motorcyclists whom I meet to imagine themselves riding in two

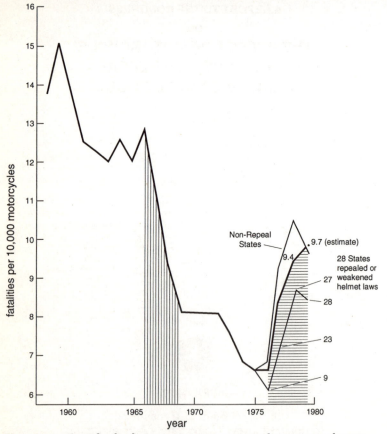

Figure 8.8 Data for fatalities per 10,000 motorcycles registered 1975–9 has been disaggregated into states that repealed their helmet laws and those that did not (*source:* Adams 1985).

different sets of clothes: first wearing only sandals, shorts and a tee shirt; secondly wearing heavy boots, leathers, and a helmet with a visor. Without exception they have agreed that the less protection they were wearing the more slowly and carefully they would ride. Although I am aware of no formally collected survey evidence with which to confirm my informal findings concerning motorcyclists' sensitivity to their perceived vulnerability, there is however considerable survey evidence that shows that fear is the primary reason that people give for not cycling at all (BMA 1992: 44). Not cycling at all as a response to the perceived danger, like not driving at all when the roads are slippery, might be labelled "gross risk compensation". It is not unreasonable to suppose that more subtle changes in behaviour, in the form of heightened vigilance and slower speeds, might occur in response to perceived differences in the consequences of an accident related to differences in the amount of crash protection worn by motorcyclists.

Alcohol and ignorance

Another area where governments of motorized countries commonly intervene is drunken driving. The claims made for these interventions rarely withstand close scrutiny. In 1983 the journal *Accident Analysis and Prevention* devoted an entire issue to the problem of impaired driving. The guest editor (Vingilis 1983) summarized his long experience of drunken driving countermeasures in a despairing introduction:

> Once again, drinking and driving has come to the fore as a public concern. The beginning of every decade over the past 30 years has seen a surge of interest in, and concern over, drinking and driving. This concern has led to millions being spent throughout the world on countermeasures, with little measurable success in reducing the problem.

The main weapons deployed in campaigns against drinking and driving are random breath testing and the *"per se* law", which deems it an offence to drive with blood alcohol concentrations above a certain level. Such measures are like speed limits in that the behaviour (or in this case the physical state) deemed criminal is not necessarily dangerous; it is associated with an increased *probability* of an accident. Figure 8.9 shows that defining criminality in terms of blood alcohol level is a rather blunt instrument. Figure 8.9a presents estimates of the relationship between blood alcohol level and the probability of having an accident. It shows that both the average probability, and the range about the average increase markedly above the legal limit of 80 mg/100 ml. Figure 8.9b suggests that much of this variability might be accounted for by age and experience.

Laws that make blood alcohol concentrations above a given level a crime *per se* are unusual in that they depend for their "success" on ignorance. The

Blennerhasset Report (DoE 1976) noted that people often sought guidance about how much they could drink without breaking the law. The report thought that such curiosity should be discouraged.

> Such requests . . . suggest a widespread and dangerous assumption that it is safe to drink up to that level. Yet by the time he reaches it a driver's ability is virtually certain to be impaired by drink. It ought to be far more widely understood that impairment is progressive and begins to develop at a level much lower than the legal limit.

Because alcohol absorption rates vary so widely between individuals, and depend also on eating patterns and physical activity, even if a motorist were to check his alcohol level before setting off on a trip, it would not guarantee that he would remain within the law for the duration of his trip. So *per se* laws operate in a fog of twofold uncertainty. Motorists have a very hazy idea of whether or not they are breaking the law, and the law-makers have a very hazy idea about whether they would be dangerous if they were.

In Chapter 7 it was noted that there was a marked decrease in the percentage of dead drivers who had blood alcohol levels above the legal limit – from 36 per cent in 1982 to 31 per cent in 1983. This decrease reversed the increase the previous year – from 31 per cent in 1981 to 36 per cent in 1982. If the unexplained "blip" in 1982 is removed from the fatality time-series presented in Figure 7.5 the accident experience of 1983 appears unexceptional. Most of the attention of researchers has focused on the question "what deserves the credit for the decrease in fatalities in 1983?" A logically prior question is "what deserves the credit for the increase in 1982?'

As Vingilis has observed, evidence for the effectiveness of countermeasures aimed at curbing drinking and driving is hard to find. It is frequently argued that examples of *temporary* success achieved by some drink–drive "blitzes" prove that the problem could be solved by some combination of more draconian penalties and more vigorous enforcement. Scandinavia is often held up as an example of what can be achieved by draconian drink–drive legislation vigorously enforced. But Ross (1976), in a report entitled "The Scandinavian myth", has shown that the available data furnish no support for the legislative deterrence thesis. His interrupted-time-series analysis revealed no effect of the legislation on the relevant accident statistics.

Ross's analysis suggests that tough drink–drive legislation is only likely to "work" where it ratifies well established public opinion. He noted the existence of a politically powerful temperance tradition in Scandinavia. Many people considered drinking and driving a serious offence (if not a sin) before it was formally designated as such by legislators. The absence of a detectable effect of Scandinavian drink–drive laws on accident statistics at the time the laws came into effect suggests that the laws were symptomatic of a widespread concern about the problem, and that most people likely to obey such

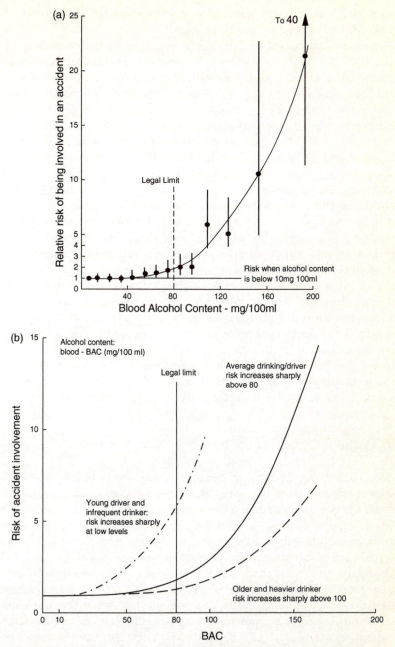

Figure 8.9 Variation in accident risk with blood alcohol content and age. (a) Variation in accident risk with blood alcohol content (*source:* DoE 1976). (b) Accident risk and blood alcohol levels (*source:* National Audit Office, *Road Safety*, June 1988, p. 12).

laws were already obeying them before they were passed. The laws, in effect, simply ratified established public opinion.

Where laws are passed that run ahead of public opinion, there appears to be a conspiracy involving motorists, the police, judges and juries to settle for a level of enforcement that accords with public opinion. In Britain since 1983 there has been a steady and impressive decrease in the number of dead drivers over the legal limit; by 1989 it was down to 19 per cent. The cause appears not to be any specific intervention by the government, but a change in social attitudes. The perceived rewards of drinking and driving have decreased, and the costs, in the form of social stigma, increased.

There remain significant exceptions to the downward trend in drinking and driving. There appear to be "high risk" and "low risk" drinkers and drivers (Adams 1988, Pelz & Schulman 1973, Wilson & Jonah 1986, Zylman 1975). For the high risk category, drinking and driving may be symptomatic of alienation and aggression which manifest themselves in a variety of anti-social behaviours. Suppressing one symptom of alienation and aggression is at best likely merely to displace anti-social behaviour, not eradicate it. A study of illicit drug use (Brown & Lawton 1988) concluded that in many instances drug use was symptomatic of rebellion against authority. The authors concluded that propaganda campaigns that emphasized the danger and illegality of drugs were more likely to encourage than deter their use. If one's objective is to spite authority, the authorities' anti-drugs advertising provides helpful guidance as to how best to do it. The Dostoevskian compulsion (discussed in Ch. 2) to assert one's individuality even at the cost of personal harm, is rarely treated seriously in the road safety literature.

The spike

In May 1993 the Department of Transport distributed half a million leaflets that purported to show that large cars are safer than small cars. Safer for whom? The leaflet contained a car safety league table that focused on the safety of drivers. But in 1991 in Britain there were about 1,700 pedestrians and cyclists killed *by* motor vehicles, considerably more than the number of drivers killed *in* motor vehicles. Studies by General Motors Research Laboratories have concluded that "driver risk-taking . . . increase[s] with increasing car mass" (Evans & Wasielewski 1983). They note that younger drivers tend to drive smaller cheaper cars, and conclude that "when the confounding effects of driver age are removed . . . we find that smaller cars have lower accident involvement rates" (Evans 1983). They also say "in response to increased perception of potential harm (that is, risk feedback), the driver of the small car reduces his risk-taking as evidenced by slower speeds, longer headways and increased seat-belt use . . . the driver of a 900kg car has 0.72 times as many crashes as a driver of a 1800kg car" (Evans 1985).

This evidence contradicts Evans's contention, reported in Chapter 7, that there is no evidence of driver responses to interventions that influence only the outcome of crashes. Indeed, elsewhere in his book Evans (1991) acknowledges very directly that the outcome of a crash is likely to affect driving behaviour.

All drivers I have questioned admit that they would drive more carefully if their vehicles contained high explosives set to detonate on impact; dramatically increasing the harm from a minor crash can clearly reduce the probability of a minor crash. I suspect that the potential embarrassment of losing my own crash-free record, which I so foolhardily announced at the beginning of this chapter, has further increased my own driving caution.

The larger your vehicle relative to everyone else on the road, the less likely you are to injure yourself and the more likely you are to injure someone else. A lorry driver in Britain is 288 times more likely than a cyclist to take some one else with him in a fatal accident. The logic of the bigger-is-safer argument, presented in the Department of Transport's leaflet comparing the "safety" of different makes of car, is the logic of the arms race; the roads, following this logic, will not be truly safe until walking and cycling have been banned and everyone is driving around in an armour-plated juggernaut.

It is primarily risk to self that governs behaviour on the road. If the principal objective of the Department of Transport were to reduce road accident casualties, the principle of risk compensation suggests a very different approach to safety. If all motor vehicles were to be fitted with long sharp spikes emerging from the centres of their steering wheels (or, if you prefer, high explosives set to detonate on impact), the disparities in vulnerability and lethality between cyclists and lorry drivers would be greatly reduced. There would probably be a redistribution of casualties, but also a reduction in the total number of casualties. Motorists driving with a heightened awareness of their own vulnerability would drive in a way that also benefited cyclists and pedestrians. Why then does the spike/explosives idea stand no chance of being adopted?

In most countries road safety is the responsibility of the same department of government that is responsible for the efficient operation of the nation's transport system. The senior politicians and civil servants who exercise these responsibilities, like individual motorists, cyclists or pedestrians, respond to the perceived rewards and costs of their actions. But the rewards and costs perceived by politicians and civil servants are very different from those experienced by people on the road. For the individual road-user the rewards of risk-taking come mainly in the form of faster journeys, and the costs mainly in the form of threats to life and limb and property. For the politicians and civil servants responsible for the regulation of transport and transport safety, the rewards and costs are very tenuously related to what happens on the road. The more senior they are, the better insulated they tend to be from the risk for which they are legislating. Their decisions, and the behaviour of their subordinates who implement their decisions, are judged by how well they conform to the demands and expectations of the hierarchy of which they are a part.

Their concerns are different. For them the pain, grief and suffering of physical accidents are reduced to statistical abstractions; indeed until a few years ago Britain's Department of Transport attached a cash value to the pain, grief and suffering of fatal accidents for the purpose of conducting cost–benefit analyses of road safety measures. It has now been reduced to a statistical variable designated "PGS" and subsumed under the "human cost" of a fatality; the value of a life at June 1991 prices is currently estimated by the Department to be £683,150 – considerably less than the £2 million to £3 million estimated by Marin and discussed in Chapter 6. The Department's estimate is made up of £237,600 worth of lost output, £450 worth of medical and ambulance costs, and £445,100 worth of "human costs" (Department of Transport 1992).

The rewards of motoring are also reduced to monetized abstractions. In their cost–benefit analyses of their road building projects, the principal benefit is time-saving for motorists. Motorists are wealthier than those without cars and their time is worth more. Most senior politicians and civil servants are motorists. People without cars are economically inferior; their concerns barely register in the cost–benefit calculations of the Department of Transport. In the formalizing of their decision-making procedures, both the convenience and the safety of people in cars are accorded greater significance than the welfare of people outside cars.

Unsupportable claims

An examination of the claims made for safety measures reveals a history of assertions, by the proponents of particular measures, that collapse under closer scrutiny. This chapter has focused on examples of this phenomenon in the realm of road safety, but Figure 4.3, showing the safety industry's failure to reduce deaths by all forms of accident and violence over a long period of time, suggests that this tendency can be found in all areas of accident prevention work.

Why? Why should those most closely involved so persistently delude themselves about their achievements? Again, cultural theory suggests some answers. In cultural theory terms, the work of planning and implementing public safety measures is the responsibility of hierarchies. Hierarchical cultures attach great importance to obedience and respect for authority. In the name of safety, governments and other large institutions require adherence to a vast range of laws and regulations. It is the defining rôle of a hierarchist to articulate and enforce codes of conduct. These codes justify the hierarchy's existence. Evidence that they are achieving something beneficial is, therefore, supportive of the hierarchist's way of life. The risk compensation hypothesis poses a threat to the hierarchy because it postulates the pervasive operation of individual autonomy. In proposing that people do not behave like obedient automatons, but that they are constantly exercising their judgement about risk, it challenges hierarchy's authority.

Most safety research is funded by hierarchies in the expectation that it will identify ways in which they might intervene to manage risk better, or demonstrate that their previous interventions have been a success. The migratory nature of risk helps to produce local safety victories that can be claimed by the managers, while the displacement effects are ignored. And the abysmal quality of most accident data, the probabilistic nature of accidents, and the fact that the probabilities are usually very small, will usually combine to reward the persistent researcher with a statistically significant result to confirm a devoutly held hypothesis. We all perforce impose meaning upon an inexplicable world.

Chapter 9

A LARGE RISK: THE GREENHOUSE EFFECT

Since the 1940s the northern half of our planet has been cooling rapidly. Already the effect in the United States is the same as if every city had been picked up by giant hands and set down more than 100 miles closer to the North Pole. If the cooling continues, warned the National Academy of Sciences in 1975, we could possibly witness the beginning of the next Great Ice Age. Conceivably, some of us might live to see huge snowfields remaining year-round in northern regions of the United States and Europe. Probably, we would see mass global famine in our lifetimes, perhaps even within a decade. Since 1970, half a million human beings in northern Africa and Asia have starved because of floods and droughts caused by the cooling climate.

This dire prospect comes from a book entitled *The cooling*, published in 1976. Since then the outlook has apparently become more threatening, but for the opposite reason; we now face the prospect of a runaway greenhouse effect. In *Global warming: the Greenpeace Report* (Leggett 1990) we are warned

... in a "business-as-usual" world in which greenhouse gas emissions continue at today's rates, we are heading for rates of temperature-rise unprecedented in human history; the geological record screams a warning to us of just how unprecedented ... And this conclusion pertains only to existing model predictions, not the natural amplifications [positive feedbacks] of global warming which the world's climate scientists profess are "likely". . .

Intriguingly, the new concern about global warming is led by some of the same scientists who were previously responsible for the concern about an impending ice age. In *The Genesis strategy* (1976) Stephen Schneider, now one of the leading advocates of international action to combat global warming, repeated the warnings of several well known climatologists that "a cooling trend has set in". By 1990 he was insisting that "the rate of change [warming] is so fast that I don't hesitate to call that kind of change potentially catastrophic for ecosystems".

Public awareness of this about-face occurred with dramatic suddenness. In *Hothouse Earth: the Greenhouse Effect and Gaia*, John Gribbin (1989) observes that "during the 1970s climatologists had become used to the idea that the world was in a *cooling* phase, retreating from the high temperatures reached in the early 1940s". He describes the circumstances in which the scientific turnaround took place.

> In 1981 it was possible to stand back and take a leisurely look at the record from 1880 to 1980 . . . . In 1987, the figures were updated to 1985, chiefly for the neatness of adding another half-decade to the records. . . . But by early 1988, even one more year's worth of data justified another publication in April, just four months after the last 1987 measurements were made, pointing out the record-breaking warmth now being reached. Even there, Hansen [James Hansen, head of the NASA team studying global temperature trends] and Lebedeff were cautious about making the connection with the greenhouse effect, merely saying that this was "a subject beyond the scope of this paper". But in the four months it had taken to get the 1987 data in print, the world had changed again; just a few *weeks* later Hansen was telling the US Senate that the first five months of 1988 had been warmer than any comparable period since 1880, and that the greenhouse effect was upon us.

Science writer Fred Pearce (1989) also captured, and contributed to, the excited atmosphere surrounding the issue in the late 1980s in his book *Turning up the heat*. Figure 9.1, redrawn from the book, shows the data for the first five months of 1988 that attracted such intense interest. The graph, ending as it does with the graph heading vertically off the top of the page, and the caption proclaiming "the greenhouse effect is here", combine to suggest a warming process rocketing out of control. The text accompanying Figure 9.1 captures the mood of eager anticipation at the time, with scientists racing to be the first with the bad news.

> But however hard the greenhouse watchers peered at their thermometers, none had been sure until 1988 that the planet was heating up as predicted. Richard Gammon of the US government's Pacific Marine Environmental Laboratory at Seattle in Washington state, seems to have been the first off the starting blocks. After seeing the complete data for 1987 and the first results from 1988, he told a conference in March 1988: "Since the mid-1970s, we have been in a period of very, very rapid warming. We are ratcheting ourselves to a new warmer climate".

Since then the world has witnessed a greenhouse-effect bandwagon-effect.

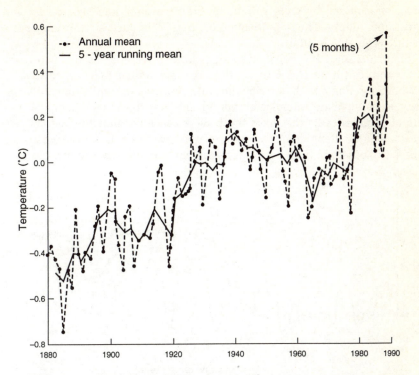

Figure 9.1 Graph from Pearce (1989). The accompanying caption reads: "A century of global warming. The picture in mid-1988, when Hansen declared 'The greenhouse effect is here'".

There has been a great rush to climb aboard a rather fragile vehicle. Intriguingly, much of the evidence for global cooling is now being recycled as evidence for global warming. Extreme or unstable *weather* conditions are adduced as evidence of changing *climate*. Ponte, in support of his cooling hypothesis, lists the following symptoms: ". . . weather gets progressively worse and tends toward extremes: heat waves and cold snaps, floods and droughts, frost and snow in the tropics and bizarre hot weather as far north as Scandinavia." Gribbin in support of the greenhouse hypothesis quotes Hansen: "The greenhouse effect is already large enough to begin to affect the probability of occurrence of extreme events such as summer heat waves . . . heatwave/drought occurrences in the Southeast and Midwest United States may be more frequent in the next decade."

Because of the year-to-year and decade-to-decade *natural* variability in global temperature, *climate* change is a process that can only be pronounced upon with any confidence on the basis of trends spanning many decades. Both the believers in cooling in the 1970s and the believers in warming in the 1990s projected historic temperature trends into the future. These trends are characterized by great variability, both temporal and geographic, and great uncertainty. The identification of "trends" in such circumstances is a highly problematic affair.

Figure 9.2, showing global mean temperatures over the past 850,000 years, is reproduced from *Hothouse Earth*; there is a similar one in *The cooling*. Gribbin and Ponte both describe the Earth, towards the end of the 20th century, as being in an "interglacial" period, and note that current temperatures are at or near levels that have not been exceeded within the last 850,000 years. The challenge confronting climatic forecasters is to say where the graph is likely to go from here. It is a problem that they share with other forecasters.

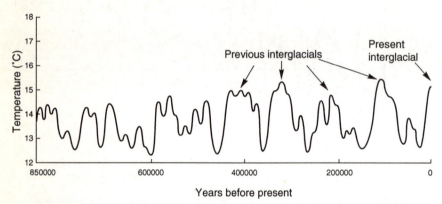

Figure 9.2 Global mean temperatures over the past 850,000 years (*source:* Gribbin 1989: 54).

Alternative futures

Figure 9.3 illustrates an exponential growth curve of the sort commonly used to describe population growth, traffic growth and economic growth in various countries in recent years. There are, to simplify somewhat, three possible ways of projecting such a trend into the future. One can assume continued exponential growth, one can assume that the process has an upper limit at which it will level off, or one can assume that at some point the graph will turn down. The first assumption is the basis of the forecasts of air traffic growth currently being used for airport planning in Britain; for the foreseeable future, growth is assumed to continue as in the past. It is also the assumption on which most economists all around the world base their middle- and long-term forecasts of gross domestic product. The second assumption, sometimes called a "saturation model" is the basis of the forecasts of car ownership used for road planning in Britain; growth is expected to stop when everyone who is old enough and fit enough to drive owns a car. It is also used

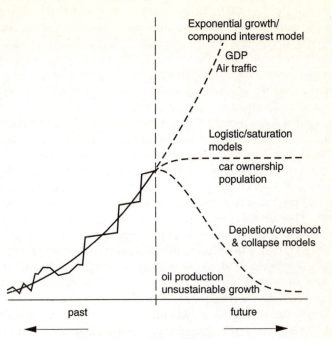

Figure 9.3   Alternative forecasts.

by population forecasters who call it the "demographic transition". The third assumption, sometimes called a "depletion curve", is used to describe the expected output of oil fields in the North Sea. It is frequently deployed in limits-to-growth debates to describe the fate of unsustainable growth processes. Each model fits the historical record equally well. The forecast is primarily determined by the choice of model, which in turn is determined by the forecaster's assumptions about the nature of the process he is attempting to predict.

The climate forecaster's problem is similar to that of the toxicologists, discussed in Chapter 3, trying to predict human responses to doses of suspect toxins for dose levels far below those tested in animal experiments. In both cases the scientists' predictions are based on extrapolation beyond the range of available data. And in both cases the extrapolations are based on assumptions about the nature of the process being predicted for which firm, uncontentious evidence is not available. But without assumptions, all such processes are simply unpredictable.

In the 1970s Ponte and a majority of climatologists looked at the record of past temperature changes and concluded that the world was near an interglacial peak. The pattern of ice ages and interglacials over the past 850,000 years suggested to scientists at that time that the graph was due to turn down, and the downward trend since 1940 was construed as evidence that it was turning down. By the 1990s, Gribbin and Pearce and many climatologists looked at the same record, plus ten more years of data, and concluded that the graph was rising, and would continue rising unless mankind reduced substantially its production of carbon dioxide and other greenhouse gases. The downward trend that had persuaded the proponents of global cooling in the 1970s that cooling was indeed under way was dismissed as evidence of "natural variability" about an upward trend.

What changed their minds? The accounts of Gribbin and Pearce quoted above suggest that it was the reversal of the cooling trend from 1940 to the early 1970s. When Ponte was writing his book in the early 1970s, temperatures in the northern hemisphere had been falling for three decades. When Gribbin was writing his book they had been rising for over a decade. The data in Figure 9.2 spanning hundreds of thousands of years have been heavily "smoothed". "Local" deviations spanning only a few centuries are averaged out to produce a relatively smooth graph. Figures 9.1 and 9.2 ought to demonstrate the futility of attempting to forecast global temperature by projecting "trends" of a few years, let alone a few months, into the future. But it appears to have been the change in recent "trends" that was the main cause of the shift in concern from cooling to warming.

There is now some doubt about whether the recent upward trend is a trend at all. The development of techniques for measuring temperatures from satellites now permits comprehensive coverage of the Earth's surface, and in particular has permitted much more extensive coverage of the oceans, and the measurement of atmospheric temperature in depth and not just at the surface. The estimates of global mean temperature shown in Figure 9.1 going back to 1880 and extended up to the present are based on thermometer measurements made at weather stations. Over time the number of these stations has increased, but, for the purpose of estimating global mean temperatures, the measurements produced by them have limitations. The stations are run by many different people with limited inspection to ensure consistent, comparable standards. Many are located in or near urban areas or airports, which are "heat islands", unrepresentative of the surrounding areas. And the coverage of the stations is sparse and uneven, with limited coverage of the southern hemisphere and very limited coverage of the oceans. The message of the surface temperature record published by the IPCC (1990) is one of a rapidly rising trend through the 1980s. But the much more comprehensive and systematic satellite measurements of global average temperature for the period 1979–93 (Fig. 9.4) reveal no trend at all. In fact the most recent data available from NASA at the time of writing reveal a very slight *downward* trend – a decadal trend over this period of –0.03°C.

Far greater uncertainty must attach to attempts such as that shown in Figure 9.2 to reconstruct the temperature record of the distant past. The further one goes back in the past, the patchier the sampling becomes and the more speculative become the methods by which past temperatures are

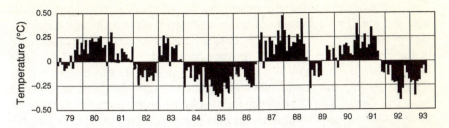

Figure 9.4 Global temperature variation in Celsius; trend since 1979, –0.03 °C per decade. All temperature variations are based on a 10-year average for the month reported. (*Source:* J. Christy, Earth System Science Laboratory, University of Alabama, Huntsville).

deduced. There is wide variability in temperature change over the Earth as average global temperatures change; Gribbin (1989) notes that during the 1980s when, according to some records, the Northern Hemisphere warmed by 0.31°C, over Scandinavia mean temperatures fell by about 0.6°C. Thus, graphs such as that displayed in Figure 9.2 of mean global temperature going back hundreds of thousands of years must be treated as speculations with a high degree of probable error.

The issue is further complicated by new arguments about whether anthropogenic CO_2 emissions are contributing to global warming at all. First, it is argued that water vapour accounts for about 97 per cent of all greenhouse gases, that CO_2 accounts for less than half the remaining 3 per cent, and that the effect of human contributions to atmospheric CO_2 through the burning of fossil fuels is too small to separate from the "noise" of natural fluctuations. Secondly, it is argued that, because cold water can hold more CO_2 than warm water, as the oceans warm they release CO_2 to the atmosphere, and it is, therefore, warming that has caused increases in atmospheric CO_2 and not the other way around. Further, it is acknowledged that CO_2 can only absorb infrared radiation at two specific narrow wavebands. It is contended by some greenhouse sceptics that these wavebands are already near saturation – that is that there is already sufficient CO_2 to absorb almost all the energy available in the relevant wavebands, and that adding further CO_2 to the atmosphere will not cause further warming because there is no more energy to be absorbed (Olstead 1993).

The debate

On 11 September 1993 the Royal Geographical Society held a mock trial in London. The prosecution framed the charge in legalistic language: "the proponents of man-made global warming are charged in that they have acted irresponsibly, causing discredit to the integrity of science and ill advised decision-making by governments without proper justification". The trial was presided over by an eminent judge, Lord Lloyd, and two Queen's Counsels presented and cross-examined witnesses who were prominent participants in the greenhouse debate.[1]The defence consisted, mostly, not of a robust justification of the scientific case for global warming, but of a review of all the caveats that the defendants had attached to their predictions of global warming. Counsel for the defence insisted in his opening statement that "we do not seek to argue that the case for man-made global warming is as yet conclusively demonstrated", and his principal witness, Dr David Carson (head of the Hadley Centre for Climate Prediction and Research) conceded that "it is simply not possible with any confidence to attribute global warming [0.5°C since 1900] to a cause such as an enhanced greenhouse effect". If the public had formed the impression that global warming was an established fact rather than a mere possibility, the responsibility, according to the defence, should be laid at the door of exaggerated reporting by the popular media.

The prosecution argued that both the climate models and the evidence from the historic record are far too primitive and unreliable to form the basis of policy recommendations, *and* that there is reason to suppose that the Earth's climate is robust and stabilized by myriad feedback mechanisms, *and* that carbon dioxide should be seen not as a pollutant, but as a fertiliser promoting plant growth. The defence accepted that conclusive proof of their fears is not yet available, *but* that there is reason to suppose that the Earth's climate is precariously balanced, and that business-as-usual is likely to lead to a runaway greenhouse effect.

Given the vehemence of some of the participants in the debate about global warming, there is a surprising degree of agreement about the inability of empirical evidence to resolve the issue. The executive summary of the 1990 report of the IPCC (Intergovernmental Panel on Climate Change) acknowledged that the increase in global temperature over the past 100 years was within the range of "natural climatic variability". Nor could they find evidence that "climates have become more variable over the last few years". And in its 1992 report the IPCC stated that "it is still not possible to attribute any or all of the warming of the last century to greenhouse gas induced climate change". This remains the view of most climatologists. A poll by *Nature* of 1500 climatologists revealed that 71 per cent thought that the climate changes of the past 100 years were "within the range of natural fluctuation" (reported by Olstead 1993).

---

1. The witnesses for the prosecution were Richard Lindzen of the Massachusetts Institute of Technology, John Emsley of Imperial College London, and Nigel Hawkes, Science Editor of The Times. The witnesses for the defence were Bert Bohlin, Chairman of the Intergovernmental Panel on Climate Change, Martin Parry from Oxford University and David Carson of the British Meteorological Office's Hadley Centre for Climate Prediction and Research.

Arguing in the dark

Human risk thermostats manifest a variable sensitivity to the behaviour of the global thermostat. The greenhouse debate turns out to be yet another case of people arguing furiously in the dark. Again the participants in the debate turn out to exhibit the biases characteristic of the stereotypes of cultural theory. The scientific disagreement about the nature of the processes at work and how to model them, and the inability of scientists to settle their arguments by appeal to empirical data, provide a fertile environment for the development of biases. Biases, like mushrooms, flourish in the dark.

The *fatalist* shrugs and smiles, amused by the exertions of those trying to make sense of an unpredictable universe. And many scientists studying climate change are rendered fatalistic by their apparent insignificance in the face of the magnitude of the processes under investigation. James Lovelock (1987) invokes the Gaia hypothesis to explain the remarkable stability of the Earth's average temperature – between 10°C and 20°C – over 3.5 billion years. But this "stability" has embraced many ice ages and greenhouses in which countless species have evolved and been extinguished. Lovelock captures the fatalistic state of mind that this perspective engenders when he observes

> People sometimes have the attitude that "Gaia will look after us". But that's wrong. If the concept means anything at all, Gaia will look after herself. And the best way for her to do that might well be to get rid of us. (Quoted in Gribbin 1989)

The *egalitarian* transfers his allegiance easily from fear of global cooling to fear of global warming. The ups and downs of the historical and geological record show that both are possible. Earth is either balanced precariously on the peak of an interglacial, ready to plunge into another ice age, or on the verge of runaway warming. His myth of nature – fragile and precarious – makes him vigilant for confirming evidence. The following passage from *The cooling* by Ponte shows that the bond uniting the "coolers" and the "warmers" is *instability*.

> Earth's climate has been cooling. This fact seems to contradict theories that say that it should be warming. But the prophets of warming *are* describing real forces that influence climate, and like other scientists are still learning how these forces interact to produce a balance of heating and cooling on our planet. It may well turn out that the growing instability of the Earth's climate is caused by human influences adding both heating *and* cooling forces to the balance, thereby making it more and more "unnatural" and precarious. The prophets of both warming and cooling agree on at least one thing: climatic changes can come quickly, within centuries or even decades, and can have devastating conse-

quences for humankind. Climatology has ceased to be a drab science. Its findings have taken on an urgent importance for all of us.

The egalitarian precautionary principle transforms uncertainty into cause for urgent action. The Greenpeace report acknowledges that "the world's climate scientists conclude in their IPCC report that "the unequivocal detection of the enhanced greenhouse effect from observations is not likely for a decade or more." But those who cannot appreciate the urgency are nevertheless viewed with disdain. The report continues:

> Ten years from now – possibly amid environmental refugees, nations on the verge of conflict over dwindling water supplies, and hosts of agricultural pests surviving increasingly warm winters – we are still likely to see scientists having difficulty putting their hands on their hearts and saying "The impacts of the enhanced greenhouse effect are now definitely being felt."

An egalitarian response to the threat of global warming is set out in a 10-point agenda in the conclusion to *Global warming: the Greenpeace Report*. It is called "Some anti-greenhouse actions for the concerned citizen"; it advocates energy-saving lightbulbs and energy-efficient appliances, draught proofing, insulating, recycling, reducing dependence on the car, organic farming, vegetarian diets, and campaigning "for anti-greenhouse changes in society" – in brief, it enjoins us all to *tread more lightly on the Earth*.

The ultimate justification for the egalitarian agenda is the possibility of a *runaway greenhouse effect*. George Woodwell (1990), president and director of the Woods Hole Research Center, demonstrates the precautionary principle in action.

> The possibility exists that the warming will proceed to the point where biotic releases [of greenhouse gases] from the warming will exceed in magnitude those controlled directly by human activity. If so, the warming will be beyond control by any steps now considered reasonable. We do not know how far we are from that point because we do not know sufficient detail about the circulation of carbon among the pools of the carbon cycle. We are not going to be able to resolve those questions definitively soon. Meanwhile, the concentration of heat-trapping gases in the atmosphere rises . . . If the process [of stabilizing the composition of the atmosphere] is not undertaken, the erosion of the human habitat will proceed rapidly, with the full panoply of ecological and political consequences.

He begins with a *possibility*, proceeds by an *if*, via doubts about *how soon*, to the *imperative for urgent action*.

Individualists, emboldened by their belief in a robust and benign nature, read the record of the past very differently; nature is benevolent and nothing has happened in the past 100 years that cannot be accounted for by "natural variability". William Nordhaus, author of the first cost–benefit analysis of the greenhouse effect (discussed below), puts the case against the precautionary principle succinctly: "To defend against the worst case will quickly bankrupt any imaginative government" (Nordhaus 1992). The precautionary principle, the sceptics observe, is indiscriminate; it has also been used to justify Star Wars and the arms race – causes to which most of those invoking it in the global warming debate would not subscribe. Human ingenuity and the march of compound interest, they say, have seen off the Malthusian prophets of doom for the past 200 years. The abrupt switch from alarm about global cooling to alarm about global warming they offer as reason to suppose that the believers in the greenhouse effect are simply alarmist.

There are scientists to serve all the established myths of nature. Individualists are reassured by the views of Richard Lindzen, Professor of Dynamic Meteorology at the Massachusetts Institute of Technology and a prominent greenhouse sceptic, who asserts that "the evidence of billions of years" proclaims the "immense robustness" of the Earth's atmosphere (cited in Olstead 1993). He protests that the models of the proponents of global warming embody *substantial positive feedback*, that is they assume global warming to be a self-amplifying process which, once started, will run on to catastrophic heating of the Earth. On the contrary, he insists, *negative feedbacks* are the norm in long-surviving stable systems, which he believes the Earth's atmosphere to be. Available empirical evidence cannot resolve the dispute. The disagreement is rooted in opposed myths of nature. Where the egalitarians present uncertainty as grounds for precautionary action, individualists find the severe limits on the present understanding of global climate grounds for optimism, and are reassured by the failure of present climate models to produce backward projections that fit the observed data. Each side points triumphantly to the inability of the other side to *disprove* the other's case. The same climatic record that renders the fatalist fatalistic, and the egalitarian fearful, renders the individualist cheerful.

Earlier false alarms are revisited. They dig up predictions of global energy shortages made during the 1970s oil crisis, and note that they have been confounded; the price of oil is now back to its pre-crisis level. A robust and benign physical nature is complemented in their outlook by an ingenious and adaptable human nature. If natural resources run short, substitutes will be invented. If the atmosphere warms up and sea levels rise, air conditioning can be installed and dykes can be built. The wealthier a nation is, the more resources it can command in the unlikely event that they might be needed to meet some future climatic challenge. The egalitarian's prescription of self-denial and reduced consumption would, they argue, slow, or stop, economic growth – the very process that increases mankind's adaptive capacity. The individualist favours business-as-usual.

Hierarchists look at the climatic record and see cause for concern but not panic. They bring the same scientific/managerial approach to the threat of global warming that they bring to all risks. What distinguishes the greenhouse effect from most other risks that they seek to manage is the unprecedented scale of the management problems it poses.

During the Cold War, climate control was seen as an issue of military significance. Lowell Ponte, the author of *The cooling*, worked during the 1960s for the US Defense Department on strategies for climate modification. In his book he rehearses some of the ideas that both the Americans and Russians had already contemplated, such as damming the Bering Straits, or using aircraft to cover vast areas of northern Russia with black dust to reduce reflectivity and increase heat absorption. The ideas now being discussed in international forums to counteract global warming – from carbon taxes and tradable CO_2 pollution permits, to schemes for massive reforestation in the Third World – *all assume that the problem being addressed is a manageable one.*

Hierarchists favour a constrained version of the precautionary principle and more research to devise effective management strategies. The British Government established the Hadley Centre for Climate Prediction and Research in 1990, with an annual budget of £12 million, to investigate global warming. But meanwhile it has set in train a wide range of precautionary research – on issues ranging from agricultural practices and alternative sources of energy to energy-saving forms of transport and land use – that is predicated on the assumption that man-made global warming is an established fact.

Of all the work that has been launched on the back of this assumption the most ambitious by far is that of the economists who are attempting a cost–benefit analysis of global warming. Economists are now being employed by the Intergovernmental Panel on Climate Change. They are insisting that a rational response to climate change requires that all costs and benefits of global warming and the control of global warming be expressed in monetary terms. A global cost–benefit analysis faces all the same problems of monetary evaluation that are encountered by the Department of Transport in doing a cost–benefit analysis of a village bypass. The most significant of these problems were discussed in Chapter 6, but the unprecedented scale of attempts to apply cost–benefit analysis to global warming merits further comment. It is the largest example I can offer of the application of the *hierarchist* approach to risk management. I call it "Vogon economics".

Vogon economics and the hyperspatial bypass

The hitchhiker's guide to the Galaxy by Douglas Adams begins with a tale of two bypasses. Both threaten the house of Arthur Dent, the bemused character at the centre of the story. A bypass presents a classic problem for economists. If built it will produce *benefits*, usually in the form of time-savings for motorists, and relief from traffic in the locality bypassed. It will also impose *costs*; it will take land and often a few houses, and bring the disturbance of traffic to a previously tranquil area.

Enter the cost–benefit analyst. His job is straightforward. He weighs up the costs and benefits, and if the latter exceed the former he concludes that there is a case for building the bypass. Cost–benefit analysis is the British Treasury's test of value for money. It is, as we have seen in Chapter 6, the quintessential hierarchist management tool. Over the past two decades almost all the new roads built in Britain, including those planned through Twyford Down and Oxleas Wood[2] have passed this test. Despite the apparent simplicity of the method and its application over many years, cost–benefit analysis remains an unsatisfactory way of resolving disputes about bypasses – unsatisfactory in the sense that those who do not want the bypass are rarely persuaded by a cost–benefit analysis that it should be built. The main difficulty is that those who enjoy the benefits of a bypass and those who bear the costs are seldom the same people. The losers are not often content with the knowledge that other people will gain more than they will lose. And attempts to compensate the losers from the gains of the winners routinely founder on disagreements about the valuation of the losses.

How, for example, should the loss of Arthur Dent's home be valued? A real-estate agent could be given the job of establishing its current market value. But what if he does not want to move? He could be paid additional compensation for his *consumer surplus*, the economist's term for the extra-market value that he places on his house. How much? According to the rules of cost–benefit analysis, the loser's loss must be the sum of money that would leave him feeling as well off after he has lost his home as before.

Unfortunately, the only person who can calculate the compensation required to leave a person feeling as well off after the event as before is the loser himself. If one's home is demolished, the geographical centre of one's existence must be relocated. For some this experience will be more upsetting than for others. For many, surveys have repeatedly confirmed, the disruption of their web of friendships and the loss of cherished surroundings cannot be compensated for by any sum of money. Many people resist the idea that such losses can be translated into cash at all, however large the sum. They either refuse to play the economist's game and decline to name a sum that would compensate them, or they say it is priceless – an answer that the economist is obliged either to disregard or to enter into his spreadsheet as infinity. As already observed in Chapter 6, it takes only one infinity to blow up a whole cost–benefit analysis.

This valuation problem is encountered not just with homes. Other build-

---

2. On 7 July 1993 the Government withdrew its proposal to build a road through Oxleas Wood in response to enormous public pressure, not as a result of a new improved cost–benefit analysis. The same method that was used to justify the Oxleas Wood scheme continues to be applied to the rest of Britain's road programme.

ings and landscapes with nostalgic associations, endangered species, security, health, and life itself also present intractable obstacles to analysts whose method requires that *everything* relevant to the decision they are trying to make should be reduced to cash.

Of the two bypasses threatening Arthur Dent's home, one was a common-or-garden local bypass of the kind that Britain's Department of Transport routinely justifies with cost–benefit analysis. The second was a Galactic Hyperspatial Express Route that required the demolition not just of Arthur Dent's home but planet Earth as well. Although the *Hitchhiker's guide* fails to say whether cost–benefit analysis was used by the Alpha Centauri planners in deciding to route their bypass through Earth, it is obvious that it must have been. In all other respects the practices of the Alpha Centauri planners and their earthling counterparts are identical. They differ only in the scale of their activities. Throughout the Galaxy it seems planners (*hierarchists*) react to protesters in the path of their projects with the same dismissive irritation. In his last announcement, before energizing the demolition beams that vaporized Earth the head of the Vogon Constructor Fleet explained that

> . . . all the planning charts and demolition orders have been on display in your local planning department in Alpha Centauri for fifty of your Earth years, so you've had plenty of time to lodge any formal complaint and it's too late to start making a fuss about it now.

The irritability of the planners stems directly from the unresolved valuation problem. The planners are persuaded that the benefits of their schemes outweigh the costs, but those in the path of their schemes rarely agree. Because some potential losers are incapable of assigning finite numbers of dollars or pounds to their losses, the planners cannot prove that the benefits of their schemes outweigh the costs. So, as we have seen in Chapter 6, they bypass the difficulty. Instead of asking people what amount of money would compensate them for their losses, they ask them what they would be willing to pay to prevent these losses. This has the effect of transforming all priceless valuations into finite numbers, and substantially reducing all lesser valuations. This in turn substantially improves the benefit:cost ratios of their projects, strengthening their argument for building what they wish to build.

The Department of Transport has hit upon a particularly effective version of this trick for valuing Sites of Special Scientific Interest, parks and other land subject to protection from development. They ask themselves what a purchaser would be willing to pay for the land if it were offered for sale in the open market *without planning permission for development*. They answer "virtually nothing"; the greater the "protection" enjoyed by a piece of land, the lower the value assigned to it by the Department's cost–benefit analysts. They call their version of cost–benefit analysis COBA; it selects routes for their schemes that have the highest benefit:cost ratios; it actively seeks out routes through the *best "protected"* parts of the country.

Where those standing in the path of a project are poor and powerless, and able to pay little or nothing to fend off the threatened destruction, and where the proposers of a project are rich and powerful and willing and able to pay a lot for the benefits, cost–benefit analysis – modified to value the losses of the poor by how much they are willing to pay to prevent them – will invariably demonstrate that the project should go ahead. Throughout the Galaxy, proposers of major projects are usually richer and stronger than the objectors, and so find cost–benefit analysis a most congenial decision-making tool. Projects whose proposers are weak and poor rarely get off the drawing board.

Tomorrow the world

Flushed with their success in the road-building industry, cost–benefit analysts are now turning their attention to a Vogon-scale problem – the threat to the Earth of the greenhouse effect. In "To slow or not to slow: the economics of the greenhouse effect", William Nordhaus (1991) explains that an efficient global strategy requires that "the costs of steps to slow climate change be balanced on the margin by the benefits in reduction of damages from climate change".

In attempting a cost–benefit analysis of the greenhouse effect the economists are attempting to attach cash values to physical effects about which there is still great scientific uncertainty and dispute. For the purpose of their analysis they are obliged to make assumptions. Nordhaus *assumes* for his cost–benefit analysis that the "damage function" increases as greenhouse gases increase. He goes on to say "I have little confidence in this assumption"[3], but nevertheless proceeds to calculate that the costs of reducing greenhouse gas emissions by half would be four times greater than the benefits. He estimates the total cost of a doubling of CO_2 at a mere 1 per cent of global GDP, considerably less than one year's growth in a good year. He concludes that, like a bypass, "climate change is likely to produce a combination of gains and losses with no strong presumption of substantial net economic damages".

Some economists appear to be so anxious to play a significant rôle in the greenhouse debate, that they are prepared to assume things that they do not believe. The essence of Nordhaus's conclusion is that, even if the greenhouse damage function is increasing, it is not very important. Fankhauser & Pearce (1993), compare Nordhaus's estimate (*based on an assumption in which he has little confidence*) with those of two subsequent studies, and report a

---

3. This statement appears in an unpublished but widely circulated 1990 version of his paper but is omitted from the version published in 1991. This passage from the original is quoted because it highlights the dependence of any analysis an economist might undertake of the greenhouse effect on evidence from physical science that is either non-existent or highly contentious.

reassuring convergence on Nordhaus's view that a doubling of CO_2 would not make a serious impact on the world economy.

> Despite differences in individual damage categories, the three studies roughly agree on the overall result, with a $2 \times CO_2$ damage in the order of 1 per cent to 2 per cent of GNP. This range turns out to be surprisingly robust. Even when picking the most pessimistic figure for each damage category the total only modestly exceeds 2 per cent of GNP. Conversely it does not fall below 3/4 per cent in the most optimistic case.

In their cost–benefit analyses, the *global costs* are the expenditures incurred in slowing warming, and the *global benefits* are the damage avoided. But *if* the IPCC scientists and other proponents of the global warming theory are right in their contention that the principal cause of global warming is the increase in CO_2 emissions caused by deforestation and the burning of fossil fuels, then the potential damage of global warming is a cost directly attributable to the growth of economic activity in industrialized countries. Most of the benefits of this activity have been enjoyed by the wealthy and powerful; most of the costs have been borne by the poor and the weak. And *if* the proponents are also right in their assumption that a substantial rise in sea level will accompany global warming, then a fair cost–benefit analysis of the greenhouse effect would have to ascertain the sum of money that would compensate the inhabitants of large parts of Bangladesh for the loss of their homes and livelihoods, and the inhabitants of small island states for the loss of their countries. The method would encounter the same difficulty discussed in Chapter 6; many of these people would answer that no sum of money could adequately compensate them for their losses, and such answers render cost–benefit analyses inoperable.

Fankhauser & Pearce bypass this problem in the same way as the Department of Transport, and with similar, although potentially much larger, effect. In their formulation of the problem, costs become benefits and benefits costs. The mighty juggernaut of economic growth, out of whose tailpipe comes the threat of global warming, is treated as an irresistible force of nature. The costs in their analysis are not the damage that the juggernaut might do to the Bangladeshis and islanders, but the costs of *preventing* this damage. The benefits are "the benefits of avoided warming". In this analysis, the rich nations on board the juggernaut do not ask those whom they are about to obliterate, "What sum of money would leave you feeling as well off after we run over you as before?" They ask, in effect, "What would your country fetch if offered for sale in the open market without planning permission for development?"

They provide a fairly specific answer; for land whose existence is threatened by sea-level rise, they assume a value ranging from $2 million to $5 million per square kilometre. This would value the benefit to the low-lying island state of Tuvalu of not being wiped out at between $52 million and $130 million (in 1989 US dollars) – or between $6,000 and $15,000 per inhabitant. Meanwhile, in the "developed" world a London–New York day return by Concorde costs $7,000 and, as Nordhaus observes, we have air conditioning and can afford to build dikes.

The application of cost–benefit analysis to the greenhouse effect is breaking new ground in one further important respect that merits a brief mention: the Vogon-scale timeframe of the analysis. Economist William Cline (1991), whose work on the economics of the greenhouse effect has been fulsomely praised by *The Economist* (11 July 1992), insists that the analysis must be extended to embrace effects 250 to 300 years in the future. Cost–benefit analysis requires all future effects to be discounted, i.e. reduced to their present value at the time of the analysis. Even at a relatively low discount rate – Cline recommends 2 per cent – effects 300 years into the future become insignificant. But Cline manages to salvage a rôle for the economist by observing that "the scale of greenhouse damage is likely to grow with the scale of GNP"; assuming that GNP will continue to grow for the next 300 years, he concludes that "the scale factor could thus neutralise much of the discounting." For the neutralization effect to be complete, global wealth (in "real" terms, i.e. constant $US) and greenhouse damage 300 years hence would have to be 380 times greater than at present.

300 years ago the US dollar did not exist and most of the North American continent was still owned by the Indians. One way of appreciating the magnitude of the task that the greenhouse economists have set themselves is to imagine them transported by time machine back to 1693, and set the task of doing a cost–benefit analysis of the European conquest of North America – with the net present value of the conquest calculated in 1693 wampum.

Doubtless the greenhouse economists would argue that the cost–benefit analyses of the greenhouse effect done so far are merely illustrative of their method, and that they require improved inputs from the scientists before they can be truly useful; Nordhaus concedes that his calculations need "fine tuning". There are two defects in this defence. First, it is likely to be a long time, if ever, before the scientific evidence is conclusive. Secondly, the economists will *never* develop a workable cost–benefit analysis that will not discriminate against the losers, because they will never have a fair valuation method that will yield finite estimates of the costs of global warming; there is no system of valuation that allows the losers to be the valuers of their own losses that is proof against one or more of the losers declaring their losses to be unassuageable by finite sums of cash. Cost–benefit analysis can therefore never answer Nordhaus's question "to slow or not to slow?" It is a method for evading, not answering, a moral question: should the wealthiest continue to grow wealthier at the expense of the poorest?

A failure to build more roads to accommodate traffic growth would, the Department of Transport argues, retard economic growth in Britain. For those who equate rationality with the reduction of all concerns to cash, all projects are to be judged ultimately by their effect on gross domestic product.

Over-zealous attempts to slow the greenhouse effect would, they argue, retard the growth of gross world product; they would divert resources from other projects with higher rates of return. Where the costs and benefits of projects are measured in US dollars, the concerns of those with the most dollars loom largest. The *project* being appraised by a cost–benefit analysis of the greenhouse effect is the promotion of world economic growth. It is comparable to a bypass through a poor suburb of Alpha Centauri to accommodate the growing traffic of wealthy Vogons travelling in air-conditioned spaceships. For Vogon economists the problem is straightforward. The benefits of the project are great. The costs are negligible. The benefit:cost ratio is substantially greater than one. The project should go ahead.

There is one small problem with this comparison. Should their project run into difficulty the Vogon economists have another planet to which they can retreat.

An introspective postscript

For a wide range of debates about risks, of which the greenhouse effect is but one of the largest, there is little or no prospect of science settling the issue. We are *all*, this author included, confronted by the need to make judgements about potential risks on the basis of inadequate evidence. The experience of assembling and discussing the evidence presented here has been accompanied by much introspection. Where do *I* stand? At times there appeared a danger of the introspection inducing total paralysis. It is easy to demonstrate that people are arguing from different premises, but if *science* is incapable of forging an agreement about premises, what more can one say? This is a problem to which I return at the end of the book.

What I have done in my discussion of the application of cost–benefit analysis to the greenhouse effect is to offer the reader an example of bias (my bias) in action. My survey of the global warming debate has made me more open minded about the scientific evidence; I began as a firm believer in man-made global warming, and am now much less sure. It has made me more fatalistic; ice ages and greenhouses of the past have occurred without the assistance of mankind and doubtless will again. It has also confirmed my prejudices about cost–benefit analysis. Having vented these prejudices I conclude that I am not a hierarchist, at least not on a global scale. The hierarchist "rationality" presumes an agreement about objectives, what the economist calls an "objective function". The hierarchist inhabits a singular hierarchy, and any hierarchist's attempt, such as that of cost–benefit analysis, to formulate a rationale for action, can only work if there are common values and agreement about the hierarchy's objectives. The method encounters insurmountable problems when attempts are made to use it to resolve disputes involving more than one hierarchy and/or other cultural biases.

The attempt by economists from affluent OECD countries to apply cost–benefit analysis to a globe-spanning problem such as the greenhouse effect is, I conclude, doomed to fail. Not only will it antagonize adherents to the other three cultural biases, it will also encounter resolute opposition from many non-OECD hierarchists.

Chapter 10

THE RISK SOCIETY

> In advanced modernity the social production of *wealth* is systematically accompanied by the social production of *risks*.
>
> The dream of the class society is that everyone wants and ought to have a *share* of the pie. The utopia of the risk society is that everyone should be *spared* from poisoning.   Ulrich Beck, *Risk society*

Ulrich Beck's *Risikogesellschaft* was published in 1986 but has only been available in English as *Risk society* since 1992. It has already emerged as a significant influence on discussions of risk. Beck's perspective, like that of cultural theory, seeks an explanation of risk in the social and cultural contexts of behaviour, and exposes the limitations of quantitative risk assessment and other scientistic attempts to pin risk down with objective measurements.

Modern science and technology, Beck argues, have created a *risk society* in which success in the production of wealth has been overtaken by the production of risk. The primary concerns of "industrial" or "class society" – the creation and equitable distribution of wealth – have been replaced, he contends, by the quest for safety.

He distinguishes modern risks from older dangers by their scale and invisibility, and the need for experts to detect them.

> Excrement piles up everywhere . . . hazards in those days [the early nineteenth century] assaulted the nose or the eyes and were thus susceptible to the senses, while the risks of civilization today typically *escape perception* and are localized in the sphere of *physical and chemical formulas* (e.g. toxins in foodstuffs or the nuclear threat).

This distinction between modern "risk" and old-fashioned "danger" or "hazard" appears somewhat exaggerated. Although human excrement was manifest to the senses in 19th century cities, the risks that it posed to health frequently were not. People died of many diseases – typhoid, smallpox, tuberculosis, bubonic plague – caused by microbes invisible to the unaided senses; in the case of cholera, for example, only rudimentary modern science, in the form of speculations that invisible organisms in London's drinking water might be the cause of the epidemics, led to the (mostly engineering) measures that brought the epidemics under control. And without modern packaging and preservatives, foodstuffs contained many, if different, toxins.

Nor is the global scale of man-made threats a recent phenomenon. The mechanisms of destruction now postulated are different, but the prospect of global destruction as a consequence of human behaviour is not a new idea; the flood at the time of Noah was ascribed to the wickedness of mankind. However, although the distinction between the fears and anxieties of preindustrial societies and those of the modern world may not be as sharp as Beck suggests, it is undoubtedly the case that science and technology have created new risks that did not exist in earlier times.

Beck pre-empts the very word "risk" for the threats of the modern world.

Risk may be defined as a *systematic way of dealing with hazards and insecurities induced and introduced by modernization itself*. Risks, as opposed to older dangers, are consequences which relate to the threatening force of modernization and to its globalization of doubt.

This definition of risk as a way of *dealing with* hazards has created unnecessary misunderstanding, especially given the common usage in English of "risk" and "hazard", noted in Chapter 5, as synonyms. Beck, by way of explanation, insists that "the concept of risk is directly bound to the concept of reflexive modernization". It would perhaps be clearer if a distinction could be maintained between cause and effect. The definition of risk as a way of "dealing with" what is commonly called risk is, at least for many English readers, a confusing tautology. But what I take this definition to be alluding to is the character of risk discussed in Chapter 3. The way people deal with something is influenced by the way they perceive it, and the act of dealing with it alters it. This is how it was put in Chapter 3: "risk perceived is risk acted upon. It changes in the twinkling of an eye as the eye lights upon it." Further on in Chapter 3 it is observed that the cultural filters through which we perceive risk are formed in our experience of dealing with it. Thus, *the perception of the probability and magnitude of some future adverse event* (the most widely shared definition of risk) is shaped by our previous experience, and undergoes continuous modification as we act upon the perception. This view of risk permits us to retain the use of the word for discussion of events that pre-date the discovery of reflexive modernization, while being consistent with Beck's use of the word in discussing the contemporary human predicament.

Risk, according to Beck, is the defining characteristic of our age. In the introduction to the English translation of *Risk society*, Scott Lash & Brian Wynne observe that

Beck sees another, darker dimension to such developments [the "Enlightenment project" of modernization] and especially in the constitutive rôle assigned to science and knowledge. For Beck the consequences of scientific and industrial development are a set of risks and hazards, the likes of which we have never previously faced.

His analysis of the "risk society" is littered, from beginning to end, with references to unprecedented dangers: "new technologies balancing on the edge of catastrophe", "as yet unknown future hazards", "[risks] systematically intensified", "irreversible harm", "atomic fallout", "ecological disaster", "[threats to] *all* forms of life on this planet", "apocalyptic threat", "the Beelzebub of multiplying risks", and "the exponential growth of risks and the impossibility of escaping from them" are but a few indicators of his perception of the modern world as being in a state of crisis. In the risk society, according to Beck, "one is no longer concerned with attaining something 'good', but rather with *preventing* the worst".

Beck and cultural theory

Beck acknowledges that risks "can . . . be changed, magnified, dramatized or minimized within knowledge, and to that extent they are particularly *open to social definition and construction*". And in the process of discussing the processes of definition and construction, his personal myth of nature becomes clear; nature has been *made* precarious and ephemeral by the advance of science and technology. He is a member of the risk society, a society which he depicts as marching under the banner "I am afraid". This society also manifests the collective solidarity that cultural theory labels egalitarian. In the risk society, he observes, *"the commonality of anxiety* takes the place of the *commonality of need* . . . [and] *solidarity from anxiety* arises and becomes a political force". Beck, in effect, consigns himself to the lower right-hand corner of cultural theory's schema as set out in Figure 3.3.

He maintains that in modern society "one" has exchanged an interest in attaining the good for a concern to prevent the worst. This exchange can be described with reference to the analogy of the risk thermostat as a diminished perception of the rewards of risk-taking, and a heightened perception of the adverse consequences of risk; and his attribution of this perceptual shift to the indefinite pronoun "one" suggests that Beck is projecting his own personal myth of nature onto the whole of modern society.

But within Beck's analysis this egalitarian perspective has competitors. Most of the other participants in the process Beck calls "reflexive modernization" can also be accommodated within the categories of cultural theory. He describes the process as leading to "the formation of a centralized state power, to concentrations of capital and to an ever more tightly woven web of divisions of labour". A cultural theorist would describe the same process as a consolidation of hierarchical power; Beck's description of state and capital ruling and profiting through the control of science and technology accords well with cultural theory's characterization of the hierarchy's exercise of power.

Modernization also leads, according to Beck, to "individualization". This is a process that gives Beck's English translator some difficulty. It involves *freisetzung*, which the translator refers to as "this ambiguous phenomenon"; the closest he can come in English is "liberation". It involves the removal of the individual from status-based classes, freeing the individual from the collective conscience, and yet at the same time increasing his or her dependence on the standardization of markets, money and the law. The ambiguity in *freisetzung* with which the translator struggles appears to result from Beck's failure on occasion to make clear that modernization is a process that affects different people differently. With appropriate subdivision the process of individualization can be made to embrace cultural theory's categories of *fatalist* and *individualist*, depending on the control that a person is capable of exercising over his or her own fate. For individualists the process of modernization is liberating and empowering, for fatalists it involves being "liberated" from traditional societal support systems and being cast adrift in a capricious and indifferent world.

Beck versus Wildavsky

Beck, it might be objected, is too complex and sophisticated a theoretician to be neatly pigeon-holed by the typology of cultural theory. But the discussion above does not pretend to be a comprehensive critique of Beck's theory of reflexive modernization. It is intended only to illustrate that one of the distinguishing features of Beck's analysis – its one-sided stress on the "darker dimension" of science and knowledge, and its doom-laden view of humanity's current precarious predicament – is characteristic of a particular perspective on nature and risk. It is a perspective that he shares with a variety of environmental campaigners and others that cultural theory labels egalitarian. However, Beck's perspective is not universally shared. There remain believers in the "Enlightenment Project" – the believers in "Progress" and the benign effects of development, modernization, and science and technology. They remain a force to be reckoned with.

The exploration of the common ground between Beck's sociological approach to risk and that of cultural theory has also been presented for the purpose of comparing the perspective of Beck with that of Aaron Wildavsky, one of the begetters of cultural theory. Despite their agreement about the importance of culture and society in moulding perceptions of risk, they come to radically different conclusions about the risks that the world is facing.

Beck's analysis is overhung with dark clouds of impending doom; science and technology are running out of control, and everywhere threaten alienation, death and destruction. The same objective reality[1] generates in Aaron Wildavsky a remarkable optimism. Consider.

Overwhelming evidence shows that the economic growth and technological advance arising from market competition have in the past two centuries been accompanied by dramatic improvements in health – large increases in longevity and decreases in sickness. (Wildavsky 1988)

It also induces a sceptical response to evidence of impending doom.

Not a day goes by without charges that products of technology harm the human body and the physical environment. The very Earth itself is said to be in serious danger . . . . it could be claimed that ours is the environmental age, the time in which technology ceased to be a liberating force and became, instead, a mechanism for self-enslavement, as if the things we created were destroying us.

The claims of harm from technology, I believe, are false, mostly false, or *unproven* [my emphasis]. To justify this conclusion I will review the relationship between scientific knowledge and governmental action in a broad spectrum of claims about low-level exposure to chemicals. (Studies of global warming, acid rain, and the thinning of the ozone layer would, I think, substantiate this argument but cannot be reported here for lack of space.) (Wildavsky 1991)

In his essay "Claims of harm from technology", Wildavsky (1991) takes the reader on a tour of the major American environmental *causes célèbres* of the past two decades: Love Canal, Agent Orange, Alar, Times Beach in Missouri, bovine growth hormone, EDB (ethylene diobromide) and asbestos. He concludes that in every case the danger was either grossly exaggerated or non-existent. More widely he offers these cases as evidence of a general propensity on the part of government regulators, politicians, the media and the general public to construe evidence of environmental harm in an alarmist way. Wildavsky the *political* scientist, like Beck the sociologist, rests his case on his assessment of the evidence of physical science.

Wildavsky's essay is one of a collection entitled *Health, lifestyle and environment: countering the panic*, which was the product of a Harvard Club Conference that addressed the question "Why are the healthiest, longest lived nations on the Earth so panicked about their health?" The historical evidence, the contributors observe, indicates a positive correlation between the advance of science and technology and increases in longevity and material wellbeing. So why worry? Wildavsky and his fellow contributors to the

---

1. Perhaps, given the inescapable relativism discussed earlier, we should say the same world viewed by a Martian.

Harvard Club Conference are exemplars of individualism. They view physical nature as benign, robust and forgiving, and human nature, or at least that part of it with which they associate, as resilient and adaptable, and impatient with bureaucratic restraint. They all, by their robust optimism, consign themselves to the lower left-hand corner of Wildavksy's own cultural theory framework. Compared to Beck and other "alarmists", far more good news about the rewards of risk gets through their cultural filters, and far less bad news concerning the costs.

Beck, a professor of sociology, and Wildavsky, a professor of political science, rest their respective cases on their convictions about "objective" reality, or what cultural theory calls their myths of nature. Both accept that risk is culturally constructed, but this insight does not liberate them from culture's grip. Indeed, it would be an unsatisfactory theory that accounted for the behaviour of everyone but the theorist. Cultural theory and risk compensation suggest where to look for an explanation of Beck's and Wildavksy's very different perspectives on risk – namely, all the previous incidents and associations in their lives that formed the cultural filters through which they now view the world.

Intriguingly, although Beck and Wildavsky agree about the significance of culture and society in shaping perspectives on risk, and disagree about the magnitude and seriousness of the risks faced by the modern world, they share a considerable measure of agreement about the way ahead.

Prescriptions

Both Beck and Wildavsky conclude on a prescriptive note. And ironically, given the force with which they have argued the case for the cultural construction of risk, both conclude that what is needed is better, more critical science, and improvements in the conduct of scientific debates. Beck first.

> Only when medicine opposes medicine, nuclear physics opposes nuclear physics, human genetics opposes human genetics, or information technology opposes information technology can the future that is being brewed up in the test-tube become intelligible and evaluable for the outside world. Enabling self-criticism in all its forms is not some sort of danger, but probably the *only way* that the mistakes that would sooner or later destroy our world can be detected in advance.

Wildavsky's conclusion is similar. Although his work on cultural theory is devoted to demonstrating the cultural relativity of risk, some perceptions of risk, he argues, are more equal than others.

> . . . great differences in perception do not signify that all are equally in line with the evidence. Though each of us may perceive what we wish,

we cannot necessarily make nature comply. Because much life and treasure are involved in conflicts over risk, it is worth trying to create more knowledge and more agreement on what counts as knowledge . . . . I have yet to see or hear a public official who says that though there is no basis in fact for the fears at which certain legislation and regulation is aimed, he is going to support it anyway because the citizenry is too dumb to know otherwise. No enterprise can exist unless the people in it make acceptable arguments to one another. Why don't we try?

How is critical science to be achieved? How can we create more knowledge and agreement about what counts as knowledge? Both Beck (from the egalitarian corner) and Wildavsky (from the individualist corner) see the problem as one of liberating science from the oppressive grip of the dominant hierarchy. Their common enemy inhabits the top right-hand corner of the cultural theory typology. Both cite many examples in which big business or big government in collusion with scientists has manipulated scientific evidence for their own ends. (It happens. In Ch. 7 I cite an example, the seat belt debate, of which I have personal experience). As a corrective, Beck advocates the institutionalization of the right to criticize one's employer.

How might such a right be protected? To ensure fair play in the debates that he advocates, Beck says "we need 'strong and independent courts'" and "strong and independent media". But how could such independence be achieved and sustained in these institutions if science itself cannot manage it? Science is, after all, *the* collective enterprise above all others to proclaim objectivity as its guiding principle. Throughout the history of science it has been the fate of ideas in conflict with the prevailing paradigm to be denounced as heresy, or dismissed as stupid or nonsensical. Even if it were possible for judges and journalists to escape the cultural biases that have captured everyone else, how might they intervene in scientific debates beyond their technical competence to separate truly stupid or nonsensical ideas from those deserving of their protection?

Beck invites us to "imagine how the discussion on reducing costs in health care could be enlivened, if we possessed an effective alternative medicine with strong arguments". It is very difficult to imagine. There already exists an alternative medicine with arguments that its adherents consider strong, but to which orthodox medicine has traditionally responded with ridicule. Disputes about medical practice provide yet another example of disputants arguing from different premises. What sort of institution or institutions might be created to overcome this problem? Beck is frustratingly vague on this crucial question; the details remain to be worked out.

> For research, . . . it would be necessary to engage in *controversial and alternative* discussions on the risks of certain steps and plans in advance, and not only in intradisciplinary circles but also in *interdisciplinary partial public spheres* that would need to be created institution-

ally. Considering that this is as yet a completely unwritten page, it seems unnecessary to think in detail about the form in which this could be organized or what monitoring possibilities these interprofessional and supraprofessional agencies would be capable of carrying out.

But orthodox science prides itself on its institutional arrangements for separating the wheat from the chaff and getting at the truth. It is precisely the issue of who is qualified to have an opinion that counts, that separates orthodox science from its "alternative" rivals. Should alchemists and astrologers be permitted an equal voice with chemists and astronomers in Beck's new interdisciplinary forums for resolving scientific disputes and detecting mistakes in advance? Who is to decide?

Beck singles out the problems of engineers for special mention.

Much would be gained . . . if the regulations that make people the opinion slaves of those they work for were reduced. Then it would be possible for engineers to report on their experiences in organizations and on the risks they see and produce, or at least they would not have to forget them once they leave work. The right to criticism within professions and organizations, like the right to strike, ought to be fought for and protected in the public interest.

In 1990 Britain's Fellowship of Engineering – the engineering profession's honorific equivalent of the Royal Society – held a conference on "Warnings of preventable disasters". The subtitle of the conference was "improving guidelines for professional engineers of all disciplines in the identification and prevention of disaster". It was, in brief, a conference focused on the question of how engineers should behave in the presence of risk. The result of their deliberations is illuminating – but not encouraging for those whose hope of progress is vested in the ability of institutions to rise above self-interest to reform themselves.

Professional disaster

Some disasters, such as epidemics and famines, are slow-motion affairs. Engineering disasters, by contrast, tend to be sudden and dramatic, involving collapse, collision, fire, explosion or escape of toxic substances. Inquests into engineering disasters, such as Piper Alpha, Zeebrugge and the Kings Cross fire, inevitably discover information which, had it been acted upon in time, would have averted the disaster. For the three years before their conference a working party with representatives from the major engineering institutions pondered the advice that they should give to their members for making or receiving warnings of disaster. One might have expected the working party's considered advice to convey a sense of urgency proportionate to the likely speed and scale of engineering disasters.

One finds a different concern. The proceedings of the conference and the product of their deliberations – in the form of a set of guidelines for engineers confronting the prospect of disaster – were published in 1991 in a book entitled *Preventing disasters*. The main message of the guidelines is that the professional engineer should ensure that, when the music at the inquest stops, he is not left without a chair.

The first paper presented to the conference was by Dr Edmund Hambly (1990), the chairman of the Fellowship's working group on the guidelines. In setting out the background to their deliberations he presented the already existing Rules of Professional Conduct for Chartered Engineers. These rules require engineers to pay due regard to:
- the safety of the public
- the interests of their client or employer
- the reputation of other engineers
- the standing of the profession.

From the case studies that Hambly then proceeded to review, it is clear that these rules are often in conflict with each other. In the first case study he recounted his discovery of faults in an offshore oil platform for which he had no direct responsibility. Should he, he asked, have risked injuring the reputation of the platform designer by pursuing his concern? Should he risk damaging the interests of his clients? While wrestling with these questions he sought a second opinion. He reported

But I found then, as I have since, that it is difficult to find someone prepared to form a really independent opinion on a controversial technical problem. Anyone who does so is drawn into a responsibility for the problem. I had to decide for myself the course of action.

Then, he reported, he spent a month doing computer calculations to confirm his initial assessment before sending a formal warning to the owner of the platform. Another week passed before a meeting at which he presented his findings to the owner. He described the reservations he had had to overcome before taking action in a section of his paper entitled "Dropping a good man in it". To his relief the owner responded, at considerable financial sacrifice, by heeding the warning and taking the platform out of service. Hambly concludes the story by recording "a strong sense of admiration for the owner and the others concerned for the decisive and honourable way in which they dealt with this uninvited problem".

The impression that one gains from this account is that Good Samaritans willing to be drawn into responsibility for other people's safety problems are difficult to find in the engineering profession; that five weeks from suspicion of catastrophe to formal warning is an uncommonly brief period; that "warners" agonize about the non-safety consequences of their warnings; and that "warnees" prepared to bear the financial cost involved in doing the honourable thing are sufficiently rare to merit a strong sense of admiration.

Hambly's second case study reinforces this impression. He was concerned in this case that political and financial considerations were overriding safety concerns in a large project on which he was working. He reported that, frustrated by the bureaucracy of the project, he sent a letter of warning to the owners by registered post and then resigned. He found that several other engineers he consulted were unconcerned about how their advice was being applied. Months passed, correspondence was exchanged, no action was taken (but no catastrophe is reported either). Hambly concluded that he had taken the matter as far as he could.

The profession's nervousness about accepting responsibility for impending disasters, illustrated by Hambly's case studies, was acknowledged in another paper to the conference by Dr John Cullen, Chairman of the Health and Safety Commission and past President of the Institution of Chemical Engineers. Effective action to prevent disasters could, he accepted, be impeded by an engineer's "self-interest, personal survival, loyalty to the company or perhaps fear of the consequences of 'blowing the whistle' on his employers".

This nervousness about accepting responsibility is not countered by the guidelines; it is enshrined in them. The version of the guidelines that accompanied the papers distributed to participants in the conference carried the following health warning:

> This document has no legal authority, it is simply recommendatory of what may be good professional practice in an appropriate case. The contributing organizations do not accept any responsibility for the way this discussion document is used or ignored.

By the time the Guidelines were published, they showed signs of anxious further editing prompted by consultations with lawyers. The health warning now reads as follows:

> This document is published by the Fellowship of Engineering solely to assist professional engineers by giving guidance to such engineers about the way they discharge their professional duties in the circumstances described above. The fellowship of Engineering hereby expressly disclaims any duty of care, or any other special relationship to any third party and specifically states that it assumes no responsibility or any risk at law, howsoever arising, for any use (including the ignoring of any warning) made by any party to these Guidelines and/or any warnings issued because of the existence of these Guidelines.

The legalistic precautionary tone continues. Amongst the "ACTIONS WHICH MIGHT BE TAKEN BY A PERSON IDENTIFYING A POSSIBLE CAUSE OF DISASTER" one finds "obtain a second opinion", "review your motives", and "maintain confidentiality". The warnee is advised "Consider your position and, if appropriate, obtain advice on legal liability and implications for insurance cover in the light of the warning received". The supplementary notes to the guidelines reiterate the importance of second opinions, legal advice, professional duties and confidentiality.

Potential warners are warned:

> Disclosure of confidential information may infringe conditions of employment, which could have serious repercussions for the employment or advancement of the warner.
>
> The warner and the warnee are likely to incur expense which is not recoverable . . . . The warner must take care not to be negligent or careless in communicating the warning. The need for legal advice should be included in the matters discussed with the Charter Engineering Institution.

There is no encouragement to press the issue to a satisfactory conclusion. On the contrary the Guidelines advise on how to make a graceful exit.

> If an informal warning is not heeded, and the warner remains convinced of the seriousness of the hazard, then he may discharge his continuing obligation under his professional Code of Conduct by issuing a formal statement to the person responsible for resolving the situation in the manner shown in paragraph 4.19 [that is by sending a warning letter by recorded delivery].

The first paragraph of the Draft Guidelines describes them as "suggested actions to assist professional engineers to consider their professional responsibilities when they ... identify or are warned of potential disaster situations".

What appears to distinguish professional responsibilities from ordinary ones are the last three items in the Rules of Professional Conduct listed above by Dr Hambly: namely concern for clients, employers and the reputations of other engineers and the profession. The guidelines issued before the conference contained not one word of urgency; not one reminder that the safety of the public should override company loyalty, interests of clients and reputations of colleagues; no encouragement to act decisively, persistently or vigorously; no urging, where all else fails, to blow whistles. In the face of disaster the Professional Engineer is expected to act "circumspectly".

The guidelines were the product of three years of deliberation and consultation amongst senior members of the engineering profession. They give the impression that "the safety of the public" has been moved from the top of the list of Professional Rules of Conduct, and is in danger of dropping off the bottom. Far from being an incentive to act in the public interest, they are a vivid reminder of the dire risks involved if one is foolhardy enough to behave in a way consistent with old-fashioned civic responsibility. They suggest that the disaster that the profession most urgently seeks to prevent is damage to the reputation of the profession.

The unimportance of being right[2]

The closing of professional ranks displayed in the Fellowship of Engineering's guidelines is characteristic of hierarchical cultures. The guidelines were the product of deliberations within the "Establishment" of the engineering profession; the working party consisted entirely of eminent engineers whose names are followed by the designation "FEng" – Fellow of the Fellowship of Engineering. In publishing their Guidelines and the deliberations that preceded them, they have provided a remarkably clear view of the workings of the collective hierarchical mind and its perception of risk. Their open and unashamed preoccupation with the welfare of the profession, to the neglect of the wider public, is a clear, albeit implicit, statement of priorities.

Not only do the guidelines show concern for the profession and its reputation, but also for the promotion prospects within the hierarchy of potential warners. Although individual engineers are rarely likely to be killed or injured by the sorts of disasters envisaged by the guidelines, they are exposed to *professional* risks – risks to job, pension, career and reputation. More generally, beyond the engineering profession, there is usually little direct connection between the risks to the general public, physical or financial, associated with decisions made by governments or commercial hierarchies, and risk to the individual decision-makers within the responsible hierarchy. The people with the greatest responsibility for decisions about societal risks of a physical nature are usually the best insulated from those

---

2. An abbreviated version of this section was published as an article in New Civil Engineer (6 September 1990) under the title "Gagging the whistle blowers". It provoked an irate letter to the editor (27 September 1990) from Edmund Hambly (the chairman of the Fellowship's working party), followed by a joint letter (18 October 1990) from Hambly and myself. The joint letter reads as follows.

Despite recent public appearances to the contrary we are agreed on the following principles that should guide engineers in the face of preventable disasters.

The first concern should be the safety of the people or environment at risk. This should be considered before personal or group loyalties, financial liability, or the interests and reputations of self or others. This we consider a general moral principle applicable not only to engineers but to any citizen.

Anyone identifying a potential disaster should respond urgently, and persist until they are satisfied that the risk has been removed, or reduced to a tolerable level.

The help of friends and professional peers should be sought, as explained in the Fellowship of Engineering Guidelines, to verify the risk, decide upon appropriate action, and reinforce the action chosen.

If warnings communicated through formal channels are blocked, or if there is a deliberate cover-up, further ways of achieving effective action should be urgently explored. As a last resort the public should be informed directly by means of exposure of the relevant information in the media.

When the guidelines were published the following year, the preamble had been amended to stress the engineer's duty to protect the public and the environment and to act promptly when confronted with a potential disaster. But the recommended actions for those giving or receiving warnings of potential disasters, and the explanatory notes, still displayed greater concern for respecting confidentiality, avoiding legal and financial liability and protecting reputations, than for avoiding disasters.

risks. Prime ministers and presidents usually travel in bullet-proof cars, surrounded by security men. They rarely live down wind of dirty smokestacks. They have food tasters (or at least testers) and bunkers to retreat to if things go seriously wrong. They are like the drivers of juggernaut lorries discussed in Chapter 8 in that their careless mistakes are unlikely to result directly in physical injuries to themselves.

In hierarchies, especially at the highest levels, the distinction between the welfare of the individual and the welfare of the organization tends to become blurred – what might be termed the *l'état c'est moi* syndrome. This blurring forms part of the collective risk management system. Henderson (1977) has described one of the most important effects of this system as "the unimportance of being right". In his study of two large British Government decisions which hindsight revealed to be two massive economic miscalculations – the Concorde and Advanced Gas Cooled Reactor projects – he was impressed by the apparent failure of these miscalculations to have had any adverse impact on the careers of those responsible.

He identified four characteristics of the British administrative process that he felt explained his findings: decorum, unbalanced incentives, anonymity and secrecy. **Decorum**, according to Henderson,[3] requires an extremely careful definition of rôles, impersonality in the transaction of business, and administrative tidiness. "These attributes are so prized" he observes "that they are sometimes uncritically equated with good administration". Decorum, he argued, has the effect of dissipating responsibility – no particular individuals are responsible – and of narrowing the flow of information to sources almost certain to toe "the party line". This in turn leads to **unbalanced incentives**. He quotes a fellow civil servant: "one thing that you might think would count [in deciding promotion], but which in fact is given no attention whatever, is whether or not your advice has been any good". He goes on to apply this to his own profession of economics: "the professional standing of an economist does not greatly depend on the success of his predictions, or on the extent to which his advice and opinions have stood the test of experience . . . . we attach little or no importance to having been right".

These problems are exacerbated by **anonymity**: "not only may it not matter greatly whether or not you were right, but few will ever know – probably no one outside a small group within your own organization". Finally, **secrecy**, he argues, weakens and dilutes responsibility, obscures and distorts the past, limits the lessons to be learned from experience, and restricts the flow of information and ideas that might illuminate future choices.

More recently, another British civil service insider, Clive Ponting (1989), has provided a similar account:

Whitehall's obsession with secrecy has far-reaching effects on the way

---

3. Henderson was writing as a liberated insider, having served in the World Bank, and as Chief Economist in the Ministry of Aviation.

Britain is governed. It makes for manipulative government by a small group of "insiders", the select few who have the right of access to "official" information in Whitehall's files . . . . It reinforces the introverted attitudes and inherited values of the policy-makers, shielding them from unwanted external pressures and enabling them quietly to select their preferred sources of advice or exercise their prejudices in private . . . . it lowers the level of public debate, it distorts the rôle of the media and it seriously hampers the way Parliament works. The veil of secrecy surrounding all official business enables the government to escape from criticism and cover up errors . . . . deficiencies in policy-making, the lack of a well argued case, and bias in background papers can remain concealed. As a substitute for real information, carefully processed and presented propaganda is pumped out for public and parliamentary consumption . . . . the government can seek to suppress dissenting views and conceal the adverse impacts of its policies. The media for the most part find it easier and safer to play by Whitehall rules, allowing themselves to be "spoon fed" with lobby briefings and press releases rather than tackle the uncertainties and risks of investigative journalism.

To avoid suffocation, keep away from children

One aspect of contemporary risks about which Beck and Wildavsky would be likely to agree is that far more of them are now under the purview of the state and large corporations. Compared to earlier, technologically simpler, times far more decisions about risk have been taken over by legislators, regulators and professional safety experts. We no longer have to worry about whether a particular restaurant is "safe"; we assume that it will have been inspected by some representative of the state and closed down if it fails to meet stringent hygiene standards. Similarly, a vast range of other physical and financial risks – from children's toys, to bank deposits, to nuclear reactors – have now become the responsibility of the state. Much of this expansion of the rôle of the state has been justified by technology's production of new risks whose scale puts them beyond the possibility of individual self-regulation. It has been accompanied by a spreading professionalization of risk, by a shift, in the terminology of cultural theory, to greater dominance in the management of risk by hierarchists.

The hierarchist approach to risk relies on experts to *anticipate* risks, and on a social system that will ensure compliance with the safety standards and rules of behaviour that are devised to defend against them. Hierarchists take a top-down managerial approach to risk. This approach involves more than mere enforcement of the law. My plastic shopping bag provides a trivial but typical example. My nearest local supermarket, Marks & Spencer, now punches holes in the bottom of the plastic bags in which it packs its customers' shopping. They do this not in response to a law, but in response to the recommendation of the Royal Society for the Prevention of Accidents. The intention of this safety measure is to make it more difficult for children accidentally to suffocate themselves. My next nearest supermarket, Sainsburys, is marginally less responsible; it merely prints on the bottom of each bag the somewhat ambiguous warning "To Avoid Suffocation Keep Away From Children". These safety precautions stem from scientific research which concluded that plastic film below 38 microns in thickness would cling to the mouth during inhalation. The decisions to punch holes or print warnings were almost certainly not made by the companies' presidents, but by middle-managers implementing their firms' anticipationist risk management philosophy.

To save the expense of buying bags which can be used to suffocate children, and which coincidentally can also be used as kitchen bin liners, I must shop at Sainsburys. The assumption by Marks & Spencer, and to a lesser extent Sainsburys, of responsibility for the safety of those of their customers' children who might come into contact with their shopping bags, appears to betray a rather low opinion of their customers' capabilities and sense of responsibility as parents. Or, it might be construed as an act of corporate protection. The increase in the share of the burden of risk management borne by the state and big business has been accompanied by, and sometimes led by, an increased propensity by judges and juries to adopt an hierarchist/ anticipationist perspective on risk, and to find someone to hold responsible for every accident.

This propensity is further advanced in the USA than in Britain and has set in train a host of compensating reactions. Hierarchies are demonstrating their ability to adapt to this increasingly dangerous legal environment. The three most common forms of risk compensation are the adoption of new low-cost safety precautions (e.g. punching holes in shopping bags), the sign-posting of every conceivable danger (e.g. printing safety warnings on shopping bags), and the purchase of more insurance – and insurance companies are compensating by increasing their premiums. And some activities unable to afford the premiums – the practice of gynaecology in the USA is an oft-cited example – are simply going out of business. There are signs of a backlash in the form of increasingly vehement protest in the press and on television. But this is countered by equally vehement insistence that the proliferation of new safety regulations is largely a legitimate response to the creation by science and technology of new dangers.

Can better science resolve the debate?

Both Beck and Wildavsky appeal for better arrangements for resolving disputes about science and technology. These appeals by two of the world's most eminent risk theorists spectacularly miss the point of all their preceding argument and analysis. It is almost as if the authors have not read their own books. Risks are culturally constructed not because people prefer make-believe to facts, but because, at the point of decision, sufficient "facts" are unavailable. Wildavsky *believes* that most claims of harm from technology are false, mostly false, or *unproven*. But in his tour of the environmental *causes célèbres* referred to above, he does not *prove* that no harm has been done, he only proves that many claims are unproven. He *believes*, as a non-scientist, that his perception is more in line with the scientific evidence than that of those who are more alarmed than he is. "Proof" is notoriously elusive, and the word is usually used, wrongly, to mean "unanimous agreement". But as we have seen in earlier chapters, long-running disagreements about risk are long-running precisely because they are unresolved, and probably unresolvable, by science. Chaos theorists now purport to have demonstrated that complex, non-linear systems, such as the weather, behave in ways that are inherently unpredictable. This problem is compounded in the realm of human behaviour by the very reflexivity which is central to Beck's sociology; it precludes the possibility of science detecting mistakes in advance. Because people are constantly responding to their circumstances, and thereby constantly altering each others' risk-taking environments – as in the dance of the risk thermostats in Chapter 2 – the future is constantly being reshaped by people's perceptions of it. Science has no firm ground on which to stand.

Douglas & Wildavsky (1983) begin their book *Risk and culture* by pointing out that the human race is condemned to act in the face of uncertainty.

> Can we know the risks we face, now or in the future? No, we cannot; but yes we must act as if we do.

"As if" is ambiguous. Some act knowing that their knowledge is partial and conditional. Others, of strong belief and conviction, manage to conjure certainty out of ignorance. Yet others, those advocating a scientific approach to risk, act *as if* uncertainty is a temporary condition that can be overcome by dint of more research. They divert attention away from the question of how to act in the face of uncertainty by focusing their energies on the impossible task of removing uncertainty.

Both Beck and Wildavsky appear to crave a certainty that the physical scientists and their own theories tell them they can never have. Both are critical of the hierarchist position – big business and government regulators – because of their practice of *bad science*. Wildavsky's principal *bêtes noires* are the US Environmental Protection Agency and Food and Drug Administration; their insistence on excessively stringent standards of environmental purity and absence of side-effects are, he argues, inhibiting the nation's wealth creation potential. Beck makes the opposite complaint; the power of modern science and technology, applied by the nuclear and chemical industries, are being insufficiently controlled; they are poisoning the world for profit. This leads them in their conclusions, for diametrically opposed reasons, to retreat, from their egalitarian and individualist positions respectively, to the rationality of the hierarchist. Big business and big bureaucracy command big science – the only form of science with the resources to address the big risks that alarm Beck, and which Wildavsky believes have been grossly exaggerated.

They retreat to this rationality with the intention of reforming it. Beck continues to view the world as precarious, but with a little pocket of safety within which the human race can shelter if, with the help of good science, it identifies the limits beyond which it must not go, and if, with the help of good government, it manages to keep everyone within these bounds. Wildavsky accepts that some things are harmful and that there are limits beyond which we must not go, but believes that good science will confirm his view that the pocket of safety is much larger than Beck believes, and that it can be made larger still if science and technology are allowed their freedom to continue to produce the improvements in nutrition, health and industrial productivity that have raised material standards of living and extended lifespans in the developed world since the beginning of the industrial revolution.

Wildavsky cites many examples of excessive regulation. Beck cites many examples, sometimes the same examples, of inadequate control of dangerous processes and substances. Both offer these examples as illustrations of bad applied science, and both feel that the scientists on their side of the argument have been getting a raw deal. But the trust they place in *good science* to sort out problems of risk is misplaced. On occasion science may succeed in solving a problem by the discovery of new agreed "facts" which can serve as a basis for consensual action; the discovery of the harmful effects of thalidomide appears to be one such case. But in such cases science has simply removed the issue from the realm of risk; it has not solved the problem of how to proceed in the absence of agreed facts. The scientific approach to risk – the Royal Society approach to risk – assumes that uncertainty is a problem that can be cracked by science. Beck argues that "where everything has become controllable, the product of human efforts, the age of excuses is over". But clearly *everything* is far from controllable. Einstein, Heisenberg, chaos theory, risk compensation and reflexivity, and empirical evidence of the sort discussed in Chapter 3 – all suggest that science can but scratch at the margins of the problem.

Chapter 11

CAN WE MANAGE RISK BETTER?

Wishful thinking

"What would happen," Ulrich Beck asks (1992), "if radiation itched?" He concludes that "in that case, nuclear policy, as well as dealing with modern mega-hazards in general, would confront a completely changed situation: the objects being disputed and negotiated would be culturally perceptible". Democratic control of risk is only possible, he insists, if we can gain "the competence to make our own judgement through a culturally created perceptibility of hazard".

Beck's rhetorical question is an example of a form of wishful thinking that bedevils most debates about risk – *if only we had more information*. Radiation does not itch, at least not at the dose levels to which human populations are normally exposed, and it is, arguably, Beck's own culturally created perception of the dangers of radiation that leads him to wish that it did. If it did, we would have more information in the form of an additional symptom; it would certainly make much easier the task of convincing those who doubt the danger of low-level radiation. It is its imperceptibility to the unaided senses, the long latency periods before any possible symptoms of damage become manifest, and ignorance of the shape of the dose–response curve at low levels, that have permitted the debate about the effects of radiation to continue unresolved for so long.

Beck is concerned to democratize risk:

Only a strong, competent public debate, "armed" with scientific arguments is capable of separating the scientific wheat from the chaff and allowing the institutions for directing technology – politics and law – to re-conquer the power of their own judgement.

A similar concern is voiced by Alan Irwin (1985) in *Risk and the control of technology*.

Decisions about risk are essentially decisions about social priorities and the values by which our societies wish to be guided. To exclude the bulk of the population from these fundamental choices would be to ensure *neither* the equity *nor* the effectiveness of regulatory policies.

But how is the trick to be done? If scientists cannot agree on how to separate the wheat from the chaff, how might the scientifically untutored "bulk of the population" participate in a rational debate about risk? In *Calculated risks*, a recent book by Joseph Rodricks (1992) on the risks to human health of chemicals in the environment, the author concludes that in most cases where chemicals are suspected of posing a threat, the risks are *incalculable* – "we don't know the truth". The ignorance surveyed by Rodricks was largely confined to the realm of toxic chemicals, in which

> the results of most risk assessments – whether expressed as an estimate of extra cancer risk or an ADI [acceptable daily intake] – are scientific hypotheses that are not generally testable with any practicable epidemiological method".

But, as we have seen in earlier chapters, his agnostic conclusion also applies to most other categories of risk that feature in long-running policy disputes.

There *are* cases in which science can reduce risk by uncovering previously unknown cause and "itch" relationships. This it has done, for example, in exposing the link between asbestos and lung disease. But even here, science has succeeded only in identifying a new risk; it has not quantified uncontentiously the dose–response relationship. It has only succeeded in identifying a risk *created* by applied science; many products incorporating asbestos were themselves, by virtue of the high electrical resistance, chemical inertness and resistance to heat of asbestos, originally seen as safety products. And although a cause and effect relationship has been established, disagreements still remain about levels of safety, about whether it should be removed from buildings in which it is incorporated, and about appropriate methods for doing so. The field of medicine is replete with examples of cures of disease which have subsequently become suspected of causing others. This problem is sufficiently common for it to have acquired its own Greek label – *iatrogenic illness*.

Whether, on balance, the rate at which science creates new risks exceeds the rate at which it removes old ones, is a question that separates technophobes (who tend to be egalitarians, constantly alert for new technogenic ills), from technophiles (individualists who see technology as bountiful and liberating), from techno-agnostics (hierarchists who see benefits from new technology only if it remains under their control). All these predispositions are on display in debates about nuclear power and genetic engineering. Both are viewed by some as Frankenstein monsters (or Pandora's boxes that should never have been opened), by others as holding out the promise of unlimited cheap power and enormous medical benefits, and by still others as potentially beneficial *if* carefully managed. All these positions rest on unverifiable, unfalsifiable, belief.

There are undoubtedly cases in which the truth about substances, products and processes harmful to the environment and human health is known and deliberately covered up by those responsible. The cloak of "commercial sensitivity" or "state security" is frequently used to conceal threats as well as actual damage. At the time of writing, the channel tunnel provides a topical example; potential users of the tunnel are currently denied information on many aspects of its safety and security on these grounds. And there are other cases in which the truth is knowable, but gets distorted by campaigning pressures – the efficacy of seat belt legislation discussed in Chapter 7 is such a case. The remedy for such cases is more open government and the participation of the vigilant and critical institutions of the sort advocated by Beck and Irwin. But such cases, in which the truth is known or knowable, ought not to be confused with genuine cases of uncertainty in which *we simply don't know the truth*.

It is common in cases of genuine uncertainty for the perceptibility of hazards to be much clearer to non-scientists than to scientists. Most of us, most of the time, navigate our way through life with remarkably simplistic ideas of the threats we face. The specialist knowledge of physicists, chemists, biologists, engineers and other "safety experts" is confined to their specialisms; and the narrower and deeper the specialism is, the more often knowledge assumes the form of a heightened awareness of ignorance. For purposes of everyday coping, we all get by, scientist and layperson alike, with crude abstractions shaped by our beliefs. The hope that science will significantly alter this state of affairs is wishful thinking.

Abstractions and the fallacy of misplaced concreteness

The philosopher Alfred North Whitehead (1932) observed,

> you cannot think without abstractions; accordingly it is of the utmost importance to be vigilant in critically revising your *modes* of abstraction.

The mistake of confusing reality with one's abstractions Whitehead dubs "the fallacy of misplaced concreteness". The myths of nature that guide people's judgements about risk might be described as background abstractions. Such abstractions are also referred to as paradigms, ideologies or beliefs – the set of assumptions about reality that are formed through shared experience, that are supported by the people with whom one associates, and that routinely go unquestioned. They are culturally constructed and culturally maintained. Whenever the behaviour of nature or people fails to accord with the expectations of these abstractions, the discrepancy must be accounted for; the fallacy of misplaced concreteness is committed when the blame is placed on the reality rather than the abstraction.

Long-running debates about risks almost always turn out on close inspection to be long-running because the participants are arguing from different premises. The framework presented here for analyzing these debates is, of course, itself a set of abstractions. It owes its present shape and form to the people with whom I have been associating, personally, and in their books and articles, for some decades. It is, like all formal abstractions, a compromise. It could have been made more detailed and complex to make it more realistic, but at a price. In the physical and social sciences, the more precise and specific become any discipline's predictions, the more arcane becomes the discourse amongst the discipline's specialists. The larger, more detailed and complex a model – the less abstract the abstraction – the smaller the number of people capable of understanding it and the longer it takes for its weaknesses and limitations to be found out. The fundamental flaw of all models that seek to illuminate the human condition, no matter how grand and complex, is that they are simplistic, inescapably so. The reality that humans seek to understand is unfathomable and incomprehensibly unique.

In statistical analysis this problem is approached by searching for correlations between variables that are suspected of influencing each other. Rarely does one find perfect correlations. Where one finds relationships that are imperfect but statistically significant, it is important, as Whitehead advised, to remain vigilant – not to lose sight of those parts of reality that do *not* conform to one's statistical model.

Cultural theory itself, which maintains that there are five and only five "viable ways of life", cannot be framed as a statistically testable hypothesis (up to this point the typology has been confined to four "ways of life'; the fifth is discussed in the next section). There are no agreed units in which individuality or adherence to a collective ethos might be measured, nor units of equality; statistically the theory remains irrefutable. Adherents to each of the five of cultural theory's ways of life are defined in terms of bundles of social relations, cultural biases, and behavioural strategies. Studies have been done in which the internal consistency of the theory's definitions has been tested (Dake 1991, Dake & Thompson 1993). In these tests, individuals were categorized on the basis of some of their social, cultural and behavioural characteristics, and hypotheses were framed about how well these categorizations would conform to other indicators of social, cultural and behavioural characteristics. The result must be deemed only a partial success; statistically significant results were obtained in many cases, but the strength of the correlations was so low as to render them of limited predictive value, and in some cases the sign of the correlation was contrary to expectation.

But these tests of definitional consistency are not tests of the theory itself; they are not tests of the central contention that the number of viable ways of life must be limited to five. Indeed, it is difficult to imagine any statistical test of the theory, given the acknowledgements (Thompson et al. 1990) that individuals often participate in a variety of social settings, and that transitions are possible from one way of life to another. If people adopt the cultural categories appropriate to several different settings while they are immersed in them – behaving, say, like egalitarians at home, hierarchists at work, and individualists while playing golf – then pinning down their "ways of life" becomes extremely problematical in cases (probably most cases), where these different social contexts cannot be cleanly separated. Transitions from one state to another also create intractable measurement problems. During such a transition Dake & Thompson (1993) maintain that people or groups display a mix of the two patterns – the one they are leaving and the one they are adopting – and not a total absence of pattern. But the real world is in a state of continual flux. Dake & Thompson (1993) observe that "our endeavour in cultural classification is made more difficult by mixed cultural orientations, complex alliances and evolving world views". Is this flux contained and ordered by the five (and only five) magnetic fields of cultural theory? Or might mixed cultural orientations and alliances be deemed to be additional viable ways of life? How might viability be measured? Measurement problems preclude framing such questions as refutable statistical hypotheses. Cultural theory, like the myths of nature it embodies, remains an abstraction beyond conclusive empirical verification.

Complicating the theory – a little bit

Following Whitehead's advice involves walking a tightrope. One cannot think without abstractions; but too much abstraction makes one a simpleton, and too little leaves one hopelessly mired in the overwhelming complexity of reality. To appease those who might consider the abstractions developed thus far to be simplistic, let us explore the possibilities of elaborating the theory. It has already been suggested that the original typology might be expanded to include alliances between different ways of life – individualist believers in free markets might, for example, team up with hierarchists to ensure that contracts are legally enforceable. Figure 11.1 introduces a further modest complication into the typology of cultural theory employed in earlier chapters (illustrated by Fig. 3.3). A vertical axis has been added; this axis is labelled the insight dimension. To the extent that we are capable of detaching ourselves from the fray and rising above it, we are likely to be better able to appreciate the partiality of those below. This perspective is sometimes called by cultural theorists (Thompson et al. 1990) the perspective of the "hermit", and his way of life is labelled "autonomy"; the hermit represents the fifth "viable way of life" in the most recent versions of cultural theory. This way of life is distinguished from the other four "engaged" ways of life by its intellectual independence and lofty detachment. This perspective might induce a resigned fatalism – people are like that, and doomed to argue for ever more from different premises. Or, it might provide the other three more activist persuasions with superior insight into the behaviour of the competing ideologies, an insight that they might seek to exploit in their management of their own risk. I will return to this problem of "reflexivity" in a moment, but first let us consider another possible elaboration of cultural theory.

In Figures 3.3 and 11.1 a typology of myths of nature has been superimposed on a typology of social relations. But it will have occurred to the reader by now that this is a convenient simplification – an abstraction – whose utility depends on the degree of correspondence between the two typologies. Figure 11.2 illustrates an expansion of the combined fourfold typology that has been used thus far. It allows for the possibility that the two typologies might not coincide in the neat way that has been assumed up to this point. Indeed, in a pluralistic world with communications media that span the globe – and with everyone exposed to novels and filmed dramas remorselessly probing the tensions, complexities and contradictions of the human condition – it would be surprising if they did. Our cultural filters must cope with a cacophony of competing and conflicting messages. If all our filters were simple and perfectly efficient, we might be organizable in the tidy way postulated by cultural theory. But often, it appears, our lives are compartmentalized in ways that permit us to hold beliefs that are mutually inconsistent.

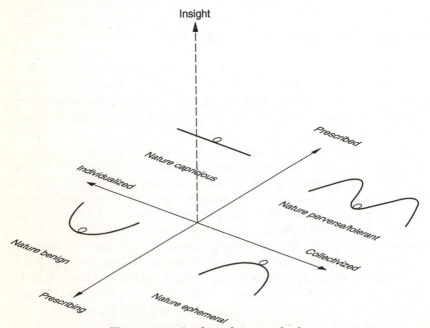

Figure 11.1   Looking down on the fray.

Figure 11.2 Cultural theory: an elaboration.

	Egalitarian	Fatalist	Hierarchist	Individualist
Ephemeral	communard Lovelock, Soros	Hell's Angel	bankrupt Lloyds "Name"	bankrupt venture capitalist
Capricious	Ned Ludd, lottery winner	Somalian refugee	BBC weather forecaster	gambler & lady luck, Lovelock
Perverse/ tolerant	Friend of Earth gov't adviser	rebel without a cause (James Dean)	ambitious civil servant, Lovelock	venture capitalist in bear market
Benign	Utopian socialist on kibbutz	Calvinist, Lovelock religious martyr	contented civil servant	venture capitalist, Soros

The version of cultural theory summarized in Figures 11.1 and 3.3 can be found in the shaded cells on the diagonal of the matrix of Figure 11.2, where egalitarians adhere to the myth of nature ephemeral, the fatalist to nature capricious, and so on. But other combinations are possible. James Lovelock (1987), the author of *Gaia*, can serve as an example. Depending on circumstances, he can be assigned to different cells in the matrix in Figure 11.2. On the geological timescale at which he contemplates the biosphere, nature is "benign" in the sense that it manifests a remarkable long-term stability; fluctuations in the Earth's atmosphere and temperature over billions of years have been confined within a range that has been hospitable to life. But these fluctuations have not always been hospitable to *human* life; they have witnessed many ice ages, and the evolution and extinction of untold numbers of species. It has already been suggested in Chapter 9 that such a perspective on the very long term, is likely to make members of the human species fatalistic; from this perspective the rôle of the human species does not appear significant.

Other evidence could lead one to classify Lovelock as an individualist. In his intellectual battles with the scientific establishment, he is a nonconformist who suffers fools badly. But he is also aware of the fickleness of "success" in the field of scientific inquiry; so might he be an individualist adherent to the view of (human) nature capricious? Sometimes, like a good egalitarian, he lends his support to environmentalists campaigning to save the whale and other endangered species. And sometimes – castigating environmentalists for exaggerating the impact of the Alaska pipeline, thereby delaying the project and costing the American economy billions of dollars – he seems a good hierarchist.

In the scientific community, there are many eminent scientists in charge of academic departments, if not whole universities, who cannot be assigned to cells on the diagonal of Figure 11.2. In their administrative lives they behave like hierarchists but they are knowledgeable about chaos theory and, on an intellectual level, one would expect them to be fatalists. The BBC's weather forecasters might also be assigned to the same cell. George Soros, the financial speculator reputed to have won a billion pounds sterling speculating against the Bank of England, when Britain was driven out of the Exchange Rate Mechanism in October 1992, is also difficult to accommodate within the simple formulation of cultural theory. As a speculator on a grand scale, he behaves like a classic individualist, behaviour difficult to reconcile with the motives underlying his philanthropic activities in eastern Europe. And what becomes of the peasant or low-paid factory worker who wins a fortune in a lottery? Some might remain fatalists and continue to see life as a lottery. Some might develop egalitarian tendencies and share their good fortune. Some might become miserly hoarders. And some might spend it quickly on high living. But all are likely to develop a new perspective on life that still retains some of the old. Established ways of life and perceptions of risk are constantly being revised in the light of new information and experience that does not conform to expectations. All the combinations represented by the cells in the matrix in Figure 11.2 are possible, but cultural theory, and the evidence considered in this book, suggests that the combinations found on the diagonal are likely to be more stable, and be representative of more people than the other combinations.

Further difficulties must be acknowledged. William Blake, whose depiction of Newton was used in Chapter 2 to illustrate the limitations of a Newtonian approach to risk, would have dismissed all of the speculations of cultural theory as hopelessly rationalistic. Blake would have been as dissatisfied with Einstein and Heisenberg as he was with Newton. The sea of time and space surrounding Newton represents the material universe that is accessible to the five senses, and accessible to science. But Blake's myth of nature floats free of such constraints, confining the choices to the possibilities represented by the four myths of nature he would have rejected as unimaginative: "to the Eyes of the Man of Imagination, Nature is Imagination itself" (quoted in Raine 1991).

A survey by *Time* magazine (27 December 1993) disclosed that 69 per cent of Americans claimed to believe in the existence of angels, and 46 per cent to believe that they have their own guardian angel. When science meets the ineffable, the concepts of quantifiable magnitude and probability that are fundamental to risk are rendered completely meaningless. Religious belief, according to the believer, has no material cause; but action motivated by religious belief does have material consequences. To the plural rationalities invoked by cultural theory to account for behaviour must be added causeless effects emanating from the ether. The best that this non-believer in angels can suggest by way of accommodating these effects is to add them to the Dance of the Risk Thermostats (Fig. 2.6) in the form of an alternative winged species that, like the Beijing Butterfly, is capable of introducing "chaos" into the system. The greater one perceives their effect to be, the more one is likely to incline to the fatalist's perception of nature as capricious – or, if one is of a religious persuasion, the more likely one is to see evidence of the deity working in mysterious ways.

The mad officials

In a book entitled *The mad officials*, Christopher Booker & Richard North (1994) present a scathing account of the work of British bureaucracy. Without using the terminology of cultural theory, they invoke, or emulate, its principal actors:

> . . . wherever we looked, and whatever type of activity we were considering, we found the great engine of bureaucracy and its myriad officials behaving in the same identifiable ways, operating to the same mindset, using the same jargon, causing the same problems, making the same mistakes. It was as if we were always looking at the same enormous, blundering monster which, although it had many heads, was always in the end the same recognizable animal. We began automatically to draw a distinction between the "world of the monster" and the "real world", where millions of people were trying to get on with their lives, run their businesses, earn their living in, on the whole, a responsible, enterprising manner. Yet wherever the monster impinged on the real world, it invariably had the same effect. It threw out clouds of deadening jargon; it tied people up in absurd paperwork and form-filling; it made ridiculous demands; it asserted its power in a blind, wilful way; it crushed enterprise and independence; at worst, it turned far too many of those who fell under its sway into nothing more than uncomprehending and often fearful victims.

The bureaucratic villains of their account are clearly hierarchists, and their "mindset" what cultural theory would label their guiding myth. The responsible, enterprising heroes of their story are individualists, and the fearful and uncomprehending victims, fatalists. For most of their book the authors themselves appear as champions of the individualist beset by the mad officials. But in their conclusion, which is focused on the perils of organophosphorous compounds, their stance shifts to that of the egalitarian, expressing concern for the welfare of society and invoking the precautionary principle: "here . . . was a perfect example of where the most stringent regulation was called for, based on proper investigation, proper science and a proper sense of care and responsibility for the health and wellbeing of enormous numbers of people".

Booker & North's description of the excesses of the bureaucrats is compelling, on occasion enraging. They present 140 specimen charges in the form of cases in which the "bureaucratic monster", in its arrogant, incompetent and arbitrary exercise of power, has put out of business, and in some cases driven to bankruptcy or suicide, egg producers, butchers, hoteliers and guest house owners, child minders, teachers, fishermen, scrap merchants and a wide variety of other innocent people engaged in socially useful enterprises. Theirs is not simply an account of the age-old battle between officaldom and the ordinary citizen; they describe a new, large, and still continuing shift in the balance of power in favour of officaldom. They note, for example, that in 1992 parliament was approving "on the nod" 10 new sets of regulations every day, that in the process of transforming the flood of new directives emanating from Brussels into UK law Whitehall was adding further requirements of its own, and that between 1991 and 1993 the number of annual inspections of food-handling premises rose from 150,000 to 419,000.

What has caused this rapid and enormous increase in regulatory activity, and why has so much of it been so damaging? It is ironic that it has taken place under a Conservative Government ostensibly committed to market forces and the liberation of the entrepreneurial spirit. The process appears to have gathered force towards the end of the 1980s economic boom. Environmentalism follows the economic cycle. At the peak of the cycle there is more traffic and more of all sorts of other activity that generates pollution or danger, and people in secure jobs are more likely to feel that they can afford the "luxury" of environmental concern. Public opinion at the start of the 1990s was therefore close to a peak in the support that it was ready to offer interventionist policies that promised to make the world safer and cleaner. The Government responded with a wide variety of commitments, most notably those embodied in its environmental White Paper, *This common inheritance*, and the Framework Convention on Climate Change that the Prime Minister signed in Rio in June 1992. At the same time the Government, spurred on by its own individualist rhetoric, was committed to shrinking the public sector. It found itself promising simultaneously more intervention and less government. Its solution was to increase greatly the productivity of those civil servants who remained by invoking the power of market forces.

But the injection of an entrepreneurial spirit into the hierarchical institutions charged with promoting health, safety and environmental protection produced an unanticipated synergy. Large numbers of quangos and Self-Financing Regulatory Agencies (SEFRAs) were established – such as the National Rivers Authority, Her Majesty's Inspectorate of Pollution, the Waste Regulatory Authorities, the Planning Inspectorate, the Fishing Vessel Survey, the Data Protection Agency and the Medicines Control Agency – and, Booker & North note, the entrepreneurs in charge of them behaved according to type. Like good businessmen they sought to bring in more business to justify the hugely increased salaries of the senior management of the newly created regulatory businesses. Increased productivity, in businesses whose principal purposes were regulation, monitoring and enforcement, meant more regulation, monitoring and enforcement. The profit motive was unleashed, but contained within the hierarchy. The capitalist growth culture was set to work in tandem with a command and control culture.

The result, the Booker & North survey suggests, is a system that combines the worst abuses of power to which each culture is prone. The capitalist is given undreamt of powers to rig the market in which his enterprise works, and the hierarchist is freed from most of the burden of democratic account-ability under which he previously operated. The risk thermostats of the risk regulators are suddenly confronted with a new environment. The rewards and risks that they seek to balance in advancing their careers bear little rela-tion to the rewards and risks facing those they are regulating. The salary and status that they seek to maximize are proportionate to their productivity – rarely measured in a way that bears any relation to the safety or productivity of those they are regulating. If crime, or food poisoning, or accidents or pol-lution damage increase, assuming the regulators can demonstrate that it is not their fault, and they usually can, this will demonstrate a need for more and better paid police and food, safety and pollution inspectors.

The Booker/North catalogue of bureaucratic cupidity and stupidity sug-gests that inviting the individualist fox into the hierarchist henhouse has had an effect exactly opposite to what was intended. But beyond implying that the traditional separation of foxes and hens provided a superior regulatory regime, it does not, unfortunately, indicate any easy remedies. They con-clude their wide-ranging tour of excessive and incompetent regulation by presenting a problem – the use of organophosphorous compounds – that they argue demonstrates the need for more stringent regulation. But on what evi-dence?

Organophosphorous is a neurotoxin in which entire sheep are routinely dipped, and survive. It produces symptoms in humans, they claim, that are "remarkably similar" to those of ME, a notoriously elusive illness. It "could be", they believe, a factor in the epidemic of suicides which had put farmers second only to vets as the most suicide-prone group in the community. An extremely fit and active person, they note, became ill "after drinking water from a hillside well, and eventually discovered that this *might have been* contaminated by OP sheep-dip used by a neighbouring farmer". The symp-toms of OP poisoning are similar to "those of the many mysterious ailments which have been so markedly on the increase in recent years . . . [such as] asthma" (on which subject see Ch. 3). When the Veterinary Medicines Direc-torate set up a system for farmers to report "*suspected* adverse reactions" from chemicals used on farms, the number of such reports increased from 63 in 1991 to 227 in 1992, and 154 related to OPs – at a time when media cover-age was inciting suspicion of OPs. No fatalities are reported by Booker & North.

Booker & North may well be right to insist on the need for more stringent regulation of OPs. But they are not toxicologists, and toxicologists are not unanimous. OPs are but one more example, to add to a multitude of others, of a risk, the perception of which is conditioned by belief. Measured on the conventional mortality scales it is a relatively minor risk. Why then, on the basis of such tentative and contentious evidence, should OPs be singled out for the application of the precautionary principle, but not most of the others? Is it, perhaps, the fact that the use of OP sheep-dip is *compulsory* that has offended the individualist instincts of Booker & North? Perhaps, but if all risks for which the evidence is no stronger than that for OPs at present were to be subjected to the same "stringent regulation" urged by Booker & North for OPs, the number of regulations and regulators needed would far exceed the present numbers that so alarm them.

So, *can* we manage risk better?

By now this question should provoke another. Who is *we*? This book was begun in the hope and expectation of being able to offer some useful advice to those seeking to manage risk. With each chapter, hopes and expectations became more modest. Cultural theory provides a typology for organizing responses to uncertainty, and each of cultural theory's types is likely to hold a distinctive view of how, or whether, risk might be managed better. The *fatalist* will answer "No, we cannot manage risk better; life is unpredictable and that is the end of it." The *individualist* will be of the opinion that there should be a devolution of managerial responsibility from bureaucracy to the individual. The *egalitarian* will favour the exercise of more caution and co-operation. The *hierarchist* will advocate more research and regulation.

Can they ever be persuaded to agree? Cultural theory suggests that it would be disastrous if they did. They each possess a small window on the truth, and help to curb the excesses of the other three. An unrestrained bureaucracy would be a Kafkaesque nightmare. Unchecked egalitarianism would produce a timid grey conformity. Unbridled individualism would allow a powerful few to overrun both nature and the many. And, if we were all fatalists, life would be nasty, brutish and short. History provides many examples: the centrally planned tyrannies in which dissent is treason, the communal tyrannies in which dissent is heresy, the injustices of rampant capitalism, and the fate of the underclasses throughout history.

Might *science* come to the rescue? Some argue that management science has been stuck in the Newtonian stage of scientific development, and that now theorists of chaos and complexity armed with powerful computers have developed new post-newtonian tools for modelling "underdetermined behaviour". But cultural theory and risk compensation prompt us to ask questions about the motives of these theorists and modellers, and how they might apply their findings. Their tools have already been used to model financial markets – to enrich the modellers themselves, or out of pure academic curiosity? And if the latter, might not someone else use their findings in an attempt to beat the market? The attempt in this book to fuse risk compensation and cultural theory engenders scepticism about methods that claim to beat the market. Any method for beating the market will have a short life. However sophisticated the modelling becomes, it will still depend on basic science for its inputs when trying to explain the behaviour of physical systems, and it will still confront the problem that its output will be someone else's input. The modellers of human and physical nature are participants in the systems they are modelling, and are constrained by the limitations of science and their vast ignorance of the physical world. The tools from chaos theory and complexity theory are more sophisticated (and comprehensible to fewer people, which ought to make them attractive to hierarchists), but they cannot solve the problem of how to act in the face of uncertainty. Indeed, by making some of the players smarter, they are likely to make the game more difficult for others. In arms races – intellectual and military – life is most dangerous for those who fall behind.

Do I have *any* practical advice about how to manage risk better? On learning that I was writing a book on the subject, people frequently demanded to know what advice I would be offering. I sensed disappointment with my answers. "Entertaining debunking", said one, "but where does it get you?" So people respond to perceived changes in safety or danger. So people vary in their perceptions of the rewards and costs of risk-taking. So people argue about risk from different premises. So what?

Initially I was frustrated by my inability to provide a satisfying answer – until I appreciated the magnitude of the question. They were asking, in effect, do I have any advice for managing *life* better? Do I have any advice that will increase the rewards and reduce the costs of risk-taking? Do I have any advice for winning at the casino, or the race track, or making a killing on the stock market? Do I have any advice that will reduce your chances of being killed in an accident or catching a fatal disease? Do I have any advice that will make you rich, and powerful, and loved, and happy, and ensure your immortality? Sorry.

The advice of others

My sense of inadequacy led me to look at how the authors of other books on risk had extricated themselves from my predicament. I made a large collection of exit lines. The quotations discussed below are but a small, but representative, sample of the parting words of other authors of books about risk. Perhaps unsurprisingly I found that the conclusions to books about risk could be assigned locations in Figure 11.1.

Most, I found, could be consigned to locations toward the bottom of the insight axis. This observation is not intended to be disparaging, merely to indicate that most books on risk are written by people seeking to control events, a possibility that appears more remote the higher one rises on the insight axis. The inscription on the tomb of Karl Marx reads *"The philosophers have only interpreted the world in various ways, the point however is to change it"*. The more committed one is to changing the world, the more difficult it becomes to take a detached and tolerant view of the thoughts and deeds of those not in agreement with your objectives. Thus, at the bottom of the insight axis one finds what cultural theorists refer to as the "engaged" ways of life. Adherents to the different engaged ways of life argue with each other from different premises, but often reserve a special contempt for the philosophers who merely interpret.

The great majority of the books on risk that I found in my survey could be assigned fairly readily to the top right-hand corner of Figure 11.1. They *do* offer specific advice of the sort my interrogators were expecting – advice about everything from choosing a safe car to having safe sex. The British Medical Association (1987) guide to *Living with risk* is a good example of this category. It concludes with a very specific and practical bit of advice.

> Reliance on human behaviour, the traditional way to improve safety in traffic, is about the worst possible way to reduce public risks. The constant use of seat belts, in the front and rear seats of cars, and ensuring that children also *never* ride unrestrained by safety seats or harnesses, are simple steps which do more to reduce the risk of death or injury in a crash (of whoever's "fault") than any other single measure.

There were at last count 740 journals being published around the world devoted to occupational health and safety;[1] those in the small fraction that I have managed to sample are filled with analyses of specific risks and advice about how to manage them better. In addition there are countless manuals and books, all written on the assumption that if you follow their advice you will reduce your risk. Most risk research is devoted to refining this advice. The approach to risk of books in this group is managerial and interventionist. Like the Royal Society they treat risk as something that is objective and measurable; it is considered a subject suitable for scientific inquiry, and the literature in this category mimics the impersonal style of the literature of physical science.

This group is distinguished by its belief in the quantifiability of risk. A book on air disasters (Grayson 1989) concludes with the advice that we should relax; the odds are pretty good.

> After you are seated and informed that the aircraft is shortly about to begin its taxi, fasten your seatbelt, place both feet firmly on the floor,

---

1. I am indebted to Herbert Eisner, former editor of *Safety Science*, for this information, which comes from a survey he conducted in 1987

close your eyes for about 10 seconds – and concentrate on the odds that are so heavily in your favour of safely reaching your destination (2,000,000 to 1). Then, open your eyes, sit back, relax – and enjoy your flight.

Every accident is a demonstration of the need to heed the advice of the professional risk manager. But their advice routinely founders on their ignorance of, or rejection of, risk compensation. An understanding of human behaviour is fundamental. If people do not wish to be safer, if they do not reduce the settings of their risk thermostats, they will frustrate the efforts of risk managers who seek to make them safer than they wish to be. The evidence reviewed in this book suggests that the principal effect of their efforts is not a reduction of risk, but a redistribution. This genre is characteristic of the *hierarchist* approach to risk.

A second distinctive group, with considerably fewer publications to its credit than the first, can be assigned to the lower right-hand corner. This group is made up of those who are acutely sensitive to the distribution issues ignored by the first. Members of this group take a campaigning approach to their subject, and are concerned to redress current injustices. Davis (1993), for example, concludes *Death on the streets* by inviting readers to share his anger.

> The outrage we feel at such an astonishing show of leniency [a £200 fine and six months driving ban for killing someone in a motoring accident] is worth nothing unless we act to control not only those cases where harm is done, but also the appalling potential for danger which all motorists present to their more vulnerable fellow road users . . . . This book hopes to let them and others see their anger as justified, and show them where this anger can be directed.

Compared to the first group, whose tone of voice is objective and scientific, this group tends to be self-consciously normative. Albury & Schwarz (1982) offer the story of the miner's safety lamp, invented by Sir Humphry Davy (President of the Royal Society from 1818–25), as a parable with which to challenge the prevailing orthodoxy found in the upper right-hand corner. The Davy Lamp, which most histories of science and safety credit with saving thousands of lives, is usually described as one of the most significant safety improvements in the history of mining. But it appears to have been a classic example of a potential safety benefit consumed as a performance benefit. Because the lamp operated at a temperature below the ignition point of methane, it permitted the extension of mining into methane-rich atmospheres; the introduction of "the safety lamp" was followed by an *increase* in explosions and fatalities.

The need for self-awareness, and the impossibility of value-free science form the theme of Albury's & Schwarz's parting advice.

We live with the illusion that science and technology happens by itself. It is not true. A socialist science recognizes this basic social fact. The rest is up to us.

This group often exudes frustration. It knows what is wrong and what needs to be done, but is exasperated by the failure of the rest of the world to share its sense of urgency. *The Greenpeace Report* (Leggett 1990) on global warming concludes with the warning that paradigms must be changed.

The uniquely frustrating thing about global warming – to those many people who now see the dangers – is that the solutions are obvious. But there is no denying that enacting them will require paradigm-shifts in human behaviour.

This group, with its collectivist ethos and sense of urgency conforms well to cultural theory's *egalitarian* category.

The advice of a third group is distinguished from that of the first two by its much greater interest in the *rewards* of risk-taking. This group is notable for its optimism. It champions free markets and espouses a trial-and-error approach to life, confident that the benefits of successful trials will outweigh the failures. The final sentences of *Searching for safety* by Aaron Wildavsky (1988) suggest that his influential rôle in the development of cultural theory does not exempt him from classification by it. He clearly belongs in the lower left-hand corner.

Safety results from a process of discovery. Attempting to short-circuit this competitive, evolutionary, trial and error process by wishing the end – safety – without providing the means – decentralized search – is bound to be self-defeating. Conceiving of safety without risk is like seeking love without courting the danger of rejection.

In debates about environmental risks the question of limits to growth frequently arises. Technological optimists do not believe in their existence. Adrian Berry (1974) provides an extreme example.

Contrary to the Club of Rome's belief, there are no limits to growth. There is no reason why our global wealth, or at least the wealth of the industrial nations, should not grow at its present annual average of 3 to 5 per cent indefinitely. Even if the Earth's resources prove ultimately to be finite, those of the Solar System and of the Great Galaxy beyond are, for all practical purposes, infinite . . . . If this 3 per cent growth rate was to be maintained, on average our civilization would then be about 10 billion times richer every 730 years.

But perhaps the best terrestrial exemplar of this group is the eponymous hero of Luke Rhinehart's (1972) novel *The dice man*. He was an optimist who trusted all important decisions in his life to the dice. The novel ends with his fate hanging by a root.

One day when Luke [the Dice Man] was being chased by two FBI men with .45s he came to a cliff and leapt off, just catching the root to a wild vine twenty yards below the ridge and dangling there. Looking down, he saw fifty feet below six policemen with machine-guns, mace, tear gas canisters and two armoured cars. Just above him he saw two mice, one white and one black, beginning to gnaw at the vine to which he clung. Suddenly he saw just in front of him a cluster of luscious ripe strawberries.

"Ah," he said, "a new option."

The advice proffered by this group can be summed up by the slogan "nothing ventured, nothing gained". Its members conform well to cultural theory's *individualist* category.

Fatalist advice about how to manage risk is something of a contradiction in terms. At the bottom of the insight axis I found no fatalist books on risk. What would be the point of writing one? But as one ascends the insight axis, as this book has attempted to do, one encounters the philosophers of risk, whose interpretations of the world leave them with few illusions about the powers of individuals to control events.

John Casti (1991), for example, concludes his book, *Searching for certainty: what scientists can know about the future*, with the words of the cosmologist John Wheeler: "The only law is that there is no law."

Some are driven to despair by their insights. Norman Dixon (1987) concludes, in his book of the same title, that we are *Our own worst enemy*.

It may well be better to be dead than confronted with the daily toll of man's irrationality . . . . It may well be better to be dead than to have to go on contemplating the fearful shortcomings of a creature which built himself a world that he is quite unfitted to inhabit.

Others are amused by the futility of the efforts of risk managers to measure our fates. Nobel prize-winning physicist Richard Feynman (1988), an irreverent individualist, might be described as being a considerable way up the insight axis, but hovering over the lower left-hand corner. Here, commenting on the Challenger disaster, he explains his preference for engineering judgement over meaningless numerical probabilities: "If a guy tells me the probability of failure is 1 in 10^5, I know he's full of crap."

Some are rendered profoundly agnostic. Ian Stewart (1989) concludes his book on the mathematics of chaos as follows:

Where will the torch of chaos lead us? We cannot tell. What is the future of chaos? It lies within the darkness. For now, we must be content to have exorcised a particularly pernicious ghost. That alone is a triumph beyond measure.

An OECD scientific expert group (1990) looking at the behaviour of motorists, concludes by acknowledging the importance of the two main themes of this book – risk compensation (which they call behavioural adaptation) and cultural differences – and offering the hope that a large research programme might yield practical advice. They might be described as hovering over the top right-hand corner of Figure 11.1:

> There is a need for multinational research which will assist in understanding cultural differences which influence behavioural adaptation. It is quite possible that a safety programme will succeed in one country, but not in another because of cultural differences which influence behavioural adaptation.

The risk management advice for which I have most sympathy is found in the concluding lines of *Risk and culture* by Mary Douglas & Aaron Wildavsky:

> If the selection of risk is a matter of social organization, the management of risk is an organizational problem. Since we do not know what risk we incur, our responsibility is to create resilience in our institutions. But by choosing resilience, which depends on some degree of trust in our institutions, we betray our bias toward the centre.

And yet. The central tension between the two main themes of this book – risk compensation and cultural theory – remains. We can all to varying degrees and for different periods of time manage to be philosophical, but we are not full-time philosophers. We all act to change the world. We all act upon our insights. Both history and cultural theory suggest that the tendency for the human race to argue, and act, from different premises is an ancient and durable one – one that centuries of philosophising has not diminished.

Ian Stewart, in his book on the mathematics of chaos puts his finger on the central intractable problem.

> Because we're *part* of the universe, our efforts to predict it may interfere with what it was going to do. This kind of problem gets very hairy and I don't want to pursue what may well be an infinite regress: I don't know how a computer would function if its constituent atoms were affected by the results of its own computations.

How to manage risk

So, emboldened by the paucity of advice offered by previous books on risk that is both practical and trustworthy, I offer my own modest suggestions. The following is a summary of the advice that this book has to offer to anyone seeking to manage risk.

214

- Remember, everyone else is seeking to manage risk too.
- They are all guessing; if they knew for certain, they would not be dealing with risk.
- Their guesses are strongly influenced by their beliefs.
- Their behaviour is strongly influenced by their guesses, *and* tends to reinforce their beliefs.
- It is the behaviour of others, and the behaviour of nature, that constitute your risk environment.
- Safety interventions that do not alter people's propensity to take risks will be frustrated by responses that re-establish the level of risk with which people were originally content.
- In the absence of reductions in people's propensity to take risks, safety interventions will redistribute the burden of risk, not reduce it.
- Potential safety benefits tend to get consumed as performance benefits.
- For the foreseeable future, nature will retain most of her secrets, and science will continue to invent new risks.
- Human behaviour will always be unpredictable because it will always be responsive to human behaviour – including your behaviour.
- It will never be possible to capture "objective risk", however powerful your computer, because the computer's predictions will be used to guide behaviour intended to influence that which is predicted.
- *In the dance of the risk thermostats, the music never stops.*

REFERENCES

AAAM 1981. *Proceedings: International Symposium on Occupant Restraint*. Morton Grove, Illinois: American Association of Automotive Medicine.

Adams, J. G. U. 1977. What noise annoys? *Vol* 1, 10–11.

—1981. The efficacy of seat belt legislation: a comparative study of road accident fatality statistics from 18 countries. Occasional Paper, Department of Geography, University College London.

—1982. The efficacy of seat belt legislation. *Society of Automotive Engineers, Transactions*, 2824–38.

—1983. Public safety legislation and the risk compensation hypothesis: the example of motorcycle helmet legislation. *Environment and Planning* C 1, 193–203.

—1985a. Smeed's Law, seat belts and the emperor's new clothes. In *Human behaviour and traffic safety*, L. Evans & R. Schwing (eds), 191–257. New York: Plenum Press.

—1985. *Risk and freedom: the record of road safety regulation*. London: Transport Publishing Projects.

—1987. Smeed's Law: some further thoughts. *Traffic Engineering and Control* (February), 70–73.

—1988. Evaluating the effectiveness of road safety measures. *Traffic Engineering and Control* (June), 344–52.

—1988b. Risk homeostasis and the purpose of safety regulation. *Ergonomics* 31(4), 407–428.

—1988b. Legal restraints on drinking and driving. Paper to EUROSAFE '88, Institute of Alcohol Studies, London.

—1988c. Risk homeostasis and the purpose of safety regulation. *Ergonomics* 31, 407–428.

—1990. Unsustainable economics. *International Environmental Affairs* 2(1). 14–21.

—1991. On being economical with the environment, *Global Ecology and Biogeography Letters* 1, 161–3.

Albury, D, & J. Schwarz 1982. *Partial progress*. London: Pluto Press.

ASBC 1978. Seat belt use abroad. Washington DC: American Seat Belt Council.

Ayer, A. J. 1965. Chance. *Scientific American* 213(October).

Beck, U. 1992. *Risk society*. London: Sage.

—1992b. From industrial society to risk society: questions of survival, social structure and ecological enlightenment. *Theory, Culture and Society* 9, 97–123.

Beer, S. 1967. *Management science*. London: Aldus.

Bell, S. 1993. *Capital Killer II: still fuming over London's traffic pollution*. A London Boroughs Association (LBA) Report.

Benjamin, B. & E. Overton 1981. Prospects for mortality decline in England and Wales, *Population Trends* 23, 22–9. London: HMSO.

Berry, A. 1974. *The next ten thousand years*. London: Hodder & Stoughton.

BMA [British Medical Association] 1992. *Cycling: towards health and safety*. Oxford: Oxford University Press.

Booker, C. & R. North 1994. *The mad officials*. London: Constable.

Born, M. 1971. *The Born–Einstein letters*. London: Macmillan.

Boyle, A. J. & C. C. Wright 1984. Accident migration after remedial treatment at accident blackspots. *Traffic Engineering and Control* 25(5), 260–67.

British Medical Association, 1980, Editorial, *British Medical Journal*.9 August.

—1983. Memorandum submitted by the BMA's Board of Science and Education to the House of Commons Transport Committee, Second Special Report, *Road Safety: inquiry not completed*, appendix 37, no. 275.

British Medical Association 1987. *Living with risk*. Chichester: John Wiley.

Broughton J. & D. C. Stark 1986. *The effect of the 1983 changes to the law relating to drink/driving*. Transport and Road Research Laboratory Research Report 89, Crowthorne, Berks.

Brown, G. R. 1991. Use of traffic conflicts for near-miss reporting. In *Near-miss reporting as a safety tool*, D. A. Lucas, T. W. Van der Schaaf, A. R. Hale (eds), ch. 10. London: Butterworth.

Brown, P. 1991. Quoted in *Local Transport Today* (30 October).

Brown, C. & J. Lawton 1988. *Illicit drug use in Portsmouth and Havant: a local study of a national problem*. London: Policy Studies Institute.

Casti, J. L. 1991. *Searching for certainty: what scientists can know about the future*. London: Abacus.

Charlton, R. 1991. Quoted in "Where the buck stops", by R. Felton in *Shell World: the International Business Magazine of Royal Dutch Shell* (February).

Cline, W. R. 1991. Scientific basis for the greenhouse effect. *The Economic Journal* 101 (July), 904–919.

Dahl, R. 1986. *Boy*. London: Penguin.

Dake, K. 1991. Orienting dispositions in the perception of risk: an analysis of contemporary worldviews and cultural biases, *Journal of Cross-Cultural Psychology* 22(1), 61–82.

Dake, K. & M. Thompson 1993. The meanings of sustainable development: household strategies for managing needs and resources. In *Human ecology: crossing boundaries*, S. D. Wright, T. Dietz, R. Borden, G. Young, G. Guagnano (eds), 421–36. FortCollins, Colorado: The Society for Human Ecology.

Davis, R. 1993. *Death on the streets*. Hawes, N. Yorkshire: Leading Edge.

Department of the Environment 1976. *Drinking and driving* [known as the Blennerhassett Report]. London: HMSO.

—1976. *Transport policy: a consultation document*. London: HMSO.

—1991. *Policy appraisal and the environment: a guide for government departments*. London: HMSO.

—1991b. Letter of 23 December 1991.

Department of Transport 1986. *Accident investigation manual* [distributed by The Royal Society for the Prevention of Accidents, Birmingham].

—1990. *Safety on the move.*

—1992. *Road accident statistics Great Britain 1991.* London: HMSO.

Dixon, N. F. 1987. *Our own worst enemy.* London: Jonathan Cape.

DOT (US Department of Transportation) 1978. Safety Belt Usage. Hearings before the subcommittee on investigations and review, of the committee on public works and transportation, US House of Representatives (95–35), US Government Printing Office.

DOT (Department of Transport) 1990. *Transport statistics Great Britain 1979–1989* (tables 2.5, 2.59, 2.60). London: HMSO.

Dostoevsky, F. 1960 [translation]. Notes from the Underground. In *Three short novels by Dostoevsky.* New York: Doubleday Anchor.

Douglas, M. 1986. *Risk acceptability according to the social sciences.* London: Routledge & Kegan Paul.

Douglas, M. & A. Wildavsky 1983. *Risk and culture: an essay on the selection of technological and environmental dangers.* Berkeley: University of California Press.

Ennals, D. 1981. *Hansard.* 13 January.

Evans, L. & P. Wasielewski 1983. *Do drivers of small cars take less risk in everyday driving?* Research Publication GMR–4425, GM Research Laboratories, Warren, Michigan.

Evans, L. 1983. *Accident involvement rate and car size.* Research Publication GMR–4453, GM Research Laboratories, Warren, Michigan.

Evans, L. 1985. *Car size and safety: results from analyzing US accident data.* Research Publication GMR-5059, GM Research Laboratories, Warren, Michigan.

Evans, L. 1991. *Traffic safety and the driver.* New York: Van Nostrand Reinhold.

Fankhauser, S. & D. W. Pearce 1993. The social costs of greenhouse gas emissions. International Conference on the Economics of Climate Change, OECD, Paris, 14–16 June.

Fellowship of Engineering (1991). *Preventing disasters.* Proceedings of conference on Warnings of Preventable Disasters. London: Fellowship of Engineering.

Feynman, R. 1988. *What do you care what other people think?* London: Unwin Hyman.

Fischoff, B., S. Lichtenstein, P. Slovic, S. Derby, R. Keeney 1981. *Acceptable risk.* Cambridge: Cambridge University Press.

Grayson, D. 1989. *Terror in the skies: the inside story of the world's worst air crashes.* Tiptree, Essex: W. H. Allen.

Gribbin, J. 1989. *Hothouse Earth: the greenhouse effect and Gaia.* London: Black Swan.

Grime, G., 1979. The protection afforded by seat belts. Report SR449, Transport and Road Research Laboratory, Crowthorne, Berkshire.

Haight, F. & R. A. Olsen 1981. Pedestrian safety in the United States: some recent trends. *Accident Analysis and Prevention* **13**, 43–55.

Hair, P. E. H. 1971. Deaths from violence in Britain: a tentative survey, *Population Studies* **25**(1), 5–24.

Hall, F. J. 1933. *Theological outlines*. London: Society for the Propagation of Religious Knowledge.

Hambly, E. 1984. Risk evaluation and realism. *Civil Engineering* (May).

Hambly, E. C. 1991. Please don't shoot the messenger, unless . . ., paper for *Warnings of preventable disasters*, Fellowship of Engineering Conference, London, 6 September 1990.

Harvey A. C. & J. Durbin 1986. The effect of seat belt legislation on British road casualties: a case study in structural time series modelling. *Royal Statistical Society, Journal* (A) **149**(3), 187–227.

Hawkes, N. 1994. Asthma increases as dust mite thrives in the modern home. *The Times* (4 March).

Henderson, P. D. 1977. Two British errors. *Oxford Economic Papers* **29**, 159–205.

Hillman, M., J. G. U. Adams, J. Whitelegg 1990. *One false move . . .: a study of children's independent mobility*. London: Policy Studies Institute.

Hillman, M. 1993. *Cycle helmets: the case for and against*. London: Policy Studies Institute.

Holling, C. S. 1979. Myths of ecological stability. In *Studies in crisis management*, G. Smart & W. Stanbury (eds). Montreal: Butterworth.

—1986. The resilience of terrestrial ecosystems. In *Sustainable development of the biosphere*, W. Clark & R. Munn (eds). Cambridge: Cambridge University Press.

Holt, R., D. Boehm-Davis, K. Fitzgerald, M. Matyuf, W. Baughman, D. Littman 1991. Behavioral validation of a hazardous thought pattern instrument, *Proceedings of the Human Factors Society 35th Annual Meeting*, 77–81.

Hurst, P. M. 1979. Compulsory seat belt use: further inferences. *Accident Analysis and Prevention* **11**, 27–33.

Hutchinson, T. P. 1987. *Road accident statistics*. Adelaide: Rumsby Scientific.

IPCC (Intergovernmental Panel on Climate Change), 1990. *Scientific Assessment of Climate Change*. WMO/UNEP, Geneva, graph p.21.

Irwin, A., 1985. *Risk and the control of technology*. Manchester: Manchester University Press.

Isles, J. E. 1981. *Seat belt savings: implications of European statistics* [unpublished]. London: [UK] Department of Transport.

Janssen, W. H., 1989. *The effect of seat belt legislation on fatality rates in a number of West European countries*. Report IZF 1989–13, TNO Institute for Perception, Soesterberg.

—1991. Seat belt wearing and driving behaviour: an empirical investigation. Report IZF 1991 C–15, TNO Institute for Perception, Soesterberg.

Key, Robert 1993. Quoted in *The Observer* (15 August).

Knight, F. 1965 (1921). *Risk, uncertainty and profit*. New York: Harper & Row.

Kryter, K. D. 1970. *The effects of noise on Man*. New York: Academic Press.

Kuhn, T. H. 1962. *The structure of scientific revolutions*. Chicago: University of Chicago Press.

Leggett, J. (ed.) 1990. *Global warming: the Greenpeace Report*. Oxford: Oxford University Press.

Lenney, W., N. E. J. Wells, B. A. O'Neil 1994. Burden of paediatric asthma. *European Respiratory Review* **4**(18), 49–62.

Lovelock, J. E. 1987. *Gaia: a new look at life on Earth*. Oxford: Oxford University Press.

Lowrance, W. 1980. The nature of risk. In *Societal risk assessment: how safe is safe enough?*, R. Schwing & W. Albers (eds), 5–14. New York: Plenum.

Mackay, M. 1981. Quoted in "Belt report slammed". *Motor* (March 7).

—1982. *Reducing car crash injuries: folklore, science and promise* (Foley Memorial Lecture). London: Pedestrians Association.

—1985. Seat belt use under voluntary and mandatory conditions and its effect on casualties. *Human behavior and traffic safety*, L. Evans & R. C. Schwing (eds), 259–83. New York: Plenum Press.

Marin, A. 1992. Cost and benefits of risk reduction. Appendix to *Risk: analysis, perception and management*. London: Royal Society.

Mishan, E. J. 1988 (1971). *Cost–benefit analysis*. London: Unwin Hyman (George Allen & Unwin).

National Audit Office 1988. *Road safety*. London: HMSO.

National Research Council 1983 (reprinted 1992). *Risk assessment in the Federal Government: managing the process*. Washington DC: National Academy Press.

Nettler, G. 1974. *Explaining crime*. New York: McGraw-Hill.

Nordhaus, W. D. 1991. To slow or not to slow: the economics of the greenhouse effect. There are two papers with this title. The first, dated 5 February 1990, is the longer unpublished version of the paper. It circulated widely in environmental circles and was much discussed prior to publication in *The Economic Journal* **101**, July 1991, 920–37.

—1992. Lethal Model 2: The Limits to Growth Revisited, Brookings Papers on Economic Activity 2, 1–59.

OECD Scientific Expert Group 1990. *Behavioral adaptations to changes in the road transport system*. Paris: OECD.

Olstead, J. 1993. Global warming in the dock. *Geographical Magazine* (September), 12–16.

Pearce, D. W. 1971. *Cost–benefit analysis*. London: Macmillan.

—1992. Green Economics. *Environmental Values* **1**(1), 3–13.

—Markandya, A., Barbier, E. 1989. *Blueprint for a Green economy*. London: Earthscan.

—& Turner, K. 1990. *Economics of natural resources and the environment*. London: Harvester Wheatsheaf.

Pearce, F. 1989. *Turning up the heat: our perilous future in the global greenhouse*. London: Bodley Head.

Pelz, D. C. & S. H. Schulman 1973. Drinking, hostility and alienation in driving of young men. *Third Alcoholism Conference*, NIAAA. Washington D.C.

Ponte, L. 1976. *The cooling*. Englewood Cliffs, NJ: Prentice-Hall.

Ponting, C. 1989. *Whitehall: changing the old guard*. London: Unwin Hyman.
Proctor, S., D. Greaves, S. Graham, M. Dias 1987. Accidents involving loss of control: identification and treatment of high-risk sites. *Proceedings of the Symposium on Crash Protection and Behavioural Aspects of Single-Vehicle Accidents*, University of Birmingham, January.

Raine, K. 1991. *William Blake*. London: Thames & Hudson.
Reason, J. 1990. *Human error*. Cambridge: Cambridge University Press.
Rhinehart, L. 1972. *The dice man*. London: Grafton.
Rodericks J. V. 1992. *Calculated risks: the toxicity and human health risks of chemicals in our environment*. Cambridge: Cambridge University Press.
Rodgers, W. 1979. *Hansard* (22 March).
Royal Society for the Prevention of Accidents, 1981, *Seat belt sense*, Birmingham: RoSPA.
—1983. *Risk assessment: a study group report*. London: Royal Society.
—1992. *Risk: analysis, perception and management*. London: Royal Society.
Rumar, K., U. Berggrund, P. Jernberg, U. Ytterbom 1976. Driver reaction to a technical safety measure – studded tires. *Human Factors* **18**, 443–54.

SAE 1984. Advances in seat belt restraint systems: design, performance and usage. Society of Automotive Engineers, P–141.
Schneider, S. H. 1976. *The genesis strategy*. New York: Plenum.
—1990. Interviewed in The Greenhouse Conspiracy, Channel 4 Television (UK), 12 August.
Schwarz, M. & M. Thompson 1990. *Divided we stand: redefining politics, technology and social choice*. Hemel Hempstead: Harvester Wheatsheaf.
Smeed, R. J. 1949. Some statistical aspects of road safety research. *Royal Statistical Society, Journal* (A) cxii (Part I, series 4), 1–24.
—1968. Variations in the pattern of accident rates in different countries and their causes. *Traffic Engineering and Control* **10**(7), 364–71.
—1972. The usefulness of formulae in traffic engineering and road safety. *Accident Analysis and Prevention* **4**, 303–312.
Stewart, I. 1989. *Does God play dice? The mathematics of chaos*. Oxford: Basil Blackwell.
Summala, H. 1988. Risk control is not risk dismount: the zero-risk theory of driver behaviour and its implications. *Ergonomics* **31**(4), 491–506.

Thompson, M. 1980. The aesthetics of risk: culture or conflict. In *Societal risk assessment: how safe is safe enough?*, R. C. Schwing & W. A. Albers (eds), 273–85. New York: Plenum.
—R. Ellis, A. Wildavsky 1990. *Cultural theory*. Boulder, Colorado: Westview.
Treasury 1991. *Economic appraisal in central government: a technical guide for government departments*. London: HMSO.
Tunbridge, R. J. 1990. The long term effect of seat belt legislation on road users injury patterns. Research Report 239, Transport and Road Research Laboratory, Crowthorne, Berkshire.

Urquhart, J. & K. Heilmann 1984. *Risk watch: the odds of life*. New York: Facts on File.

van der Colk, H. 1988. Risky behaviour resulting from bounded rationality. *Ergonomics* **31**(4), 485–90.

Vingilis, E. R. 1983. Guest editor's introduction. *Accident Analysis and Prevention* **15**, 405–406.

Visser, K. 1971, First Reaction. *Shell World: the international business magazine of Royal Dutch Shell*, February 1991.

Watson, G. S., P. L. Zador, A. Wilks 1981. Helmet use, helmet use laws, and motorcyclist fatalities. *American Journal of Public Health* **71**, 297–300.

Wildavsky, A. 1988. *Searching for safety*. Oxford: Transition.

—1991. If claims of harm from technology are false, mostly false, or unproven, what does that tell us about science? In *Health, lifestyle and environment*, P. Berger et al. (eds), ch. 10. London: Social Affairs Unit/Manhattan Institute.

Wilde, G. 1976. The risk compensation theory of accident causation and its practical consequences for accident prevention. Paper presented at the annual meeting of the Österreichische Gesellschaft für Unfallchirurgies, Salzburg.

Wilson, J. J. & B. A. Jonah 1986. Impaired drivers and high-risk drivers: are they chips from the same block? *10th International Conference on Alcohol, Drugs and Safety*.

Williams A. F. & B. O'Neil 1974. On-road driving records of licenced race drivers. *Accident Analysis and Prevention* **6**, 263–270.

Whitehead, A. N. 1932. *Science and the modern world*. Cambridge: Cambridge University Press.

Woodwell, G. M. 1990. The effects of global warming. In *Global warming: the Greenpeace Report*, J. Leggett (ed.), ch. 5. Oxford: Oxford University Press. [The Woods Hole Research Centre, of which Woodwell is president and director, is a small private organization created by Woodwell, and is not to be confused with the well known Woods Hole Oceanographic Institution and the Woods Hole Marine Biological Laboratories.]

Wright, C., A. Boyle, J. Redgrove 1988. Subjective and objective risk in road accident causation: the objective risk problem. *Proceedings of the International Symposium on Traffic and Safety Theory and Research Methods*, Session on Theoretical Analysis and Models. Amsterdam: Netherlands Institute for Road Safety Research (SWOV).

Yankauer, A. 1981. Deregulation and the right to life (editorial). *American Journal of Public Health*, 797–8.

Zylman, R. 1975. Mass arrests for impaired driving may not prevent traffic deaths. *Proceedings of the 6th Conference on Alcohol, Drugs and Traffic Safety*. Toronto, September 1974.

INDEX